国家自然科学基金面上项目(41971356)资助
国家自然科学基金青年基金(41701446)资助
教育部产学合作协同育人项目(202102136009)资助

KQGIS WebGIS 软件开发实践
KQGIS WebGIS RUANJIAN KAIFA SHIJIAN

主　编：郭明强　刘星瑞　陈麒玉　赵保睿
副主编：朱　江　钟　静　容东林　黄　颖

图书在版编目(CIP)数据

KQGIS WebGIS 软件开发实践/郭明强等主编．—武汉：中国地质大学出版社，2023.6
ISBN 978-7-5625-5616-9

Ⅰ.①K… Ⅱ.①郭… Ⅲ.①地理信息系统-应用软件-程序设计 Ⅳ.①P208

中国国家版本馆 CIP 数据核字(2023)第 108734 号

KQGIS WebGIS 软件开发实践		郭明强 等主编
责任编辑：王 敏	选题策划：王 敏	责任校对：张咏梅

出版发行：中国地质大学出版社(武汉市洪山区鲁磨路 388 号)	邮政编码：430074
电 话：(027)67883511 传 真：(027)67883580	E-mail:cbb@cug.edu.cn
经 销：全国新华书店	http://cugp.cug.edu.cn
开本：787 毫米×1 092 毫米 1/16	字数：358 千字 印张：14
版次：2023 年 6 月第 1 版	印次：2023 年 6 月第 1 次印刷
印刷：武汉市籍缘印刷厂	
ISBN 978-7-5625-5616-9	定价：78.00 元

如有印装质量问题请与印刷厂联系调换

前　言

苍穹地理信息服务平台（KQGIS Server）基于高性能的跨平台 GIS 内核，支持海量空间数据免切片快速发布，支持二三维一体化的服务发布、管理与聚合，提供强大的空间大数据存储、分析、处理等多种 Web 服务，具备多层次的扩展开发能力，可快速构建基于云端一体化的 GIS 应用系统。

KQGIS 2D for Leaflet 是一套基于 Leaflet 库开发的 WebGIS 二次开发平台，支持访问 KQGIS Server 发布的地图、数据、GIS 分析功能等资源，可为用户提供专业的 GIS 功能，同时提供了优秀的可视化能力和灵活的扩展能力。平台支持主流 GIS 厂商发布的标准 OGC（open geospatial consortium，开放地理空间信息联盟）服务，支持简单高效、稳定灵活的插件扩展，支持 Chrome、FireFox、IE 等多种主流浏览器，支持多种地图和数据可视化效果。

基于 KQGIS 2D for Leaflet，KQGIS 平台可为国土、市政、电信、交通等领域构建网络地理信息行业应用系统，提供针对空间数据的可视化、查询、检索、编辑与分析等服务。本教材基于 KQGIS 2D for Leaflet 平台二次开发接口，在客户端实现各类 WebGIS 服务的调用、交互和结果展示等功能，为读者提供了从地图基础功能到目录服务，再到数据服务、要素服务、几何服务、空间分析、专题图等功能的实现流程，为读者掌握基于 KQGIS 的 WebGIS 软件开发原理和开发方法提供了较为全面的技术指导。

笔者长期从事有关网络地理信息服务的理论方法研究、教学和软件开发工作，先后编写了《WebGIS 之 OpenLayers 全面解析》《WebGIS 之 OpenLayers 全面解析（第 2 版）》《WebGIS 之 Leaflet 全面解析》《WebGIS 之 Element 前端组件开发》《WebGIS 之 ECharts 大数据图形可视化》《WebGIS 之 Cesium 三维软件开发》等书，对目前主流的 WebGIS 开发技术进行了详细介绍。笔者十余年的 WebGIS 平台相关研发经验和应用开发基础，为本教材的编写打下了扎实的知识基础。本教材由国家自然科学基金项目（41971356，41701446）和教育部产学合作协同育人项目（202102136009）资助。从实验环境部署到各个 WebGIS 功能的开发，本教材涵盖了网络地理信息服务的关键内容。内容按照由易到难的顺序讲解，循序渐进，使读者更容易掌握知识点，同时对重点代码做了注释和讲解，以便于读者更加轻松地学习。

本教材面向广大 WebGIS 开发爱好者，内容编排遵循一般学习曲线，由浅入深地介绍了基于 KQGIS 的 WebGIS 功能开发相关知识点，内容完整，实用性强，既有详尽的理论阐述，又有丰富的案例程序，使读者能快速、全面地掌握基于 KQGIS 平台的 WebGIS 开发技术。即使对初学者来说也没有任何门槛，按部就班跟着教程实例编写代码即可。无论您是否拥有 WebGIS 编程经验，都可以借助本教材来系统了解和掌握基于 KQGIS 平台 WebGIS 二次开发 API 的网络地理信息服务开发所需的技术知识点，为掌握和理解网络地理信息服务奠定良好的基础。

教材资源：

本教材提供配套的全部示例源码，每个实验对应的源码工程均是独立编写而成的。每个源码工程可以独立运行，可快速查看演示效果与完整源码，可通过微信扫描二维码下载配套数据资源与工程源码。

参与本教材编写的还有苏望发、官小平、杨春、张晖、彭宜平、曹荣龙、陈振兴、朱涛、黄俊。本教材的出版得到中国地质大学出版社的鼎力支持，在此致以诚挚的谢意。同时向教材所涉及参考资料的所有作者表示衷心的感谢，如有引用文献缺失，请反馈至出版社，我们将在下一版中进行修正。

因作者水平有限，不足之处在所难免，敬请读者批评指正。

郭明强

2023 年 3 月于武汉

目 录

1 KQGIS 概述 (1)

1.1 KQGIS 产品体系 (1)
1.2 KQGIS Desktop 平台简介 (1)
1.3 KQGIS Server 平台简介 (3)
1.4 KQGIS Mobile 平台简介 (3)
1.5 KQGIS 2D for Leaflet 简介 (4)

2 KQGIS WebGIS 开发环境配置 (5)

2.1 KQGIS Server 平台安装 (5)
2.2 KQGIS WebGIS 客户端 SDK 配置 (8)
2.3 示例数据库配置 (9)
2.4 矢量地图服务发布 (11)
2.5 二维影像服务发布 (14)
2.6 WMTS、WMS、WFS 服务发布 (19)

3 地图基础功能 (22)

3.1 地图创建与第三方地图导入 (22)
3.2 基本地图控件 (27)
3.3 地图图层切换 (33)
3.4 地图数据测量 (34)
3.5 地图信息标注 (37)
3.6 地图动画展示 (43)

4 KQGIS 目录服务 (49)

4.1 获取数据源列表及数据目录 (49)
4.2 获取已发布的地图服务信息 (53)
4.3 获取数据库中某图层信息 (61)

5 KQGIS 数据服务 ··· (65)

5.1 瓦片地图加载 ·· (65)
5.2 矢量地图文档加载 ·· (66)
5.3 矢量瓦片加载 ·· (72)
5.4 属性瓦片加载 ·· (73)
5.5 第三方地图加载 ··· (77)
5.6 OGC 地图服务加载 ··· (82)

6 KQGIS 要素服务 ··· (90)

6.1 查询要素 ·· (90)
6.2 编辑要素 ·· (105)

7 KQGIS 几何服务 ··· (131)

7.1 几何坐标投影转换 ·· (131)
7.2 几何拓扑分析 ·· (138)
7.3 要素缓冲分析 ·· (142)
7.4 多边形裁剪分析 ··· (147)
7.5 要素叠加 ·· (151)
7.6 计算要素实地周长 ·· (155)
7.7 计算要素实地面积 ·· (157)

8 KQGIS 空间分析 ··· (159)

8.1 图层坐标投影转换 ·· (159)
8.2 图层拓扑查错 ·· (167)
8.3 图层缓冲分析 ·· (170)
8.4 图层叠置分析 ·· (173)
8.5 路径分析 ·· (178)

9 KQGIS 专题图 ·· (186)

9.1 客户端专题图 ·· (186)
9.2 专题图服务 ··· (207)

主要参考文献 ··· (215)

1 KQGIS 概述

突如其来的新型冠状病毒感染打破一切常规,在带来巨大负面经济影响同时,也加快新技术在社会经济中的应用速度。在这次疫情中,与社会经济息息相关的 GIS 技术也再次展示出不可替代的作用,国产 GIS 技术替代更势不可挡!

近年来,我国地理空间技术取得突破性进展,GIS 正逐渐成为消费市场的主流。在国内,GIS 给人们带来便利,成了人们生活中不可或缺的一部分。不仅是 GIS 技术,优秀 GIS 产品的动态也同样值得期待。苍穹地理信息(KQGIS)平台是由苍穹数码技术股份有限公司完全自主研发的、通过专家评测的、国内首个全面支持国产化环境的地理信息平台。KQGIS 平台将逐年完成更新迭代,更加便捷、高效、智能的 GIS 软件产品值得期待!

1.1 KQGIS 产品体系

作为 GIS 应用的技术核心,GIS 基础软件将是现在和未来很长一段时间内 GIS 企业核心竞争力的具体表现。苍穹数码不但拥有 KQGIS 这样的 GIS 基础软件,还积累了超过 300 项 GIS 行业应用,能够为广大客户提供强大的技术支撑和成熟产品。苍穹数码坚持走 GIS 创新化道路,结合市场动态与客户需求,不断融合大数据、云计算、物联网、人工智能等 IT 前沿技术,加大与科研单位、用户单位的合作力度,以"能更简单地解决问题"为目标,让 KQGIS 的功能更加强大、使用更加简便。KQGIS V8.0 产品体系完善、功能强大,包含桌面 GIS、三维 GIS、服务 GIS、大数据 GIS 以及移动 GIS 等专业应用与二次开发包,同时还具备二三维数据整合与管理、空间大数据分析与可视化、高性能服务发布与共享以及简便型二次开发等能力。其产品体系架构如图 1-1 所示。

1.2 KQGIS Desktop 平台简介

苍穹地理信息桌面(KQGIS Desktop)平台是一款专业型桌面 GIS 应用与开发软件,具备强大的数据采集与编辑、地理分析与处理、地图制图与输出等能力;支持二三维数据一体化管理、展示、分析及服务发布;实现跨平台安装部署,突破传统桌面 GIS 软件在 Windows 环境下运行的局限性(图 1-2)。

平台特点:便捷的数据采集与编辑工具,跨平台部署运行;高级的地理分析与处理能力,二三维一体化;强大的地图制图与输出功能,服务发布与管理。

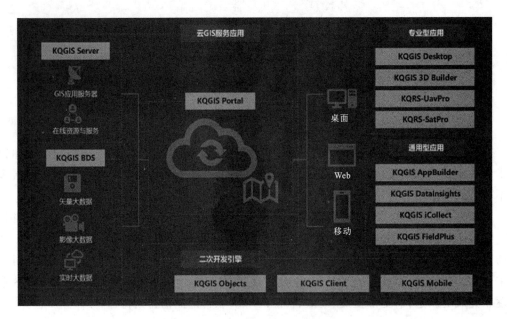

图1-1 KQGIS产品体系架构图

图1-2 KQGIS Desktop界面

1.3 KQGIS Server 平台简介

苍穹地理信息服务平台(KQGIS Server)基于高性能的跨平台 GIS 内核,支持海量空间数据免切片快速发布;支持二三维一体化的服务发布、管理与聚合;提供强大的空间大数据存储、分析、处理等多种 WEB 服务;具备多层次的扩展开发能力;可快速构建基于云端一体化的 GIS 应用系统(图 1-3)。

图 1-3 KQGIS Server REST 服务界面

平台特点:二三维一体化服务发布与管理;海量空间数据存储与分析;GIS 资源按需使用、动态伸缩、智能分配;具备服务二次开发自定义扩展能力。

1.4 KQGIS Mobile 平台简介

苍穹地理信息移动开发(KQGIS Mobile)平台支持 Android 和 IOS 操作系统的智能移动终端;具备专业而全面的 GIS 功能,支持多样化数据采集编辑、海量离/在线数据管理、流畅的地图体验、二三维数据可视化、专业的空间分析等,满足用户对移动 GIS 的应用需求;支持多种专业数据和服务,可以快速开发各类在线和离线的移动 GIS 应用。

平台特点:兼容 Android/IOS 系统,跨终端数据直接交换;数据采编方式丰富多样,离/在线地图操作流畅;移动 GIS+3D/AR/VR,GPS 和 BDS 双星定位。

1.5　KQGIS 2D for Leaflet 简介

　　KQGIS 2D for Leaflet 是一套基于 Leaflet 库开发的 WebGIS 二次开发平台，支持访问 KQGIS Server 发布的地图、数据、GIS 分析功能等资源，为用户提供了完整、专业的 GIS 能力，同时提供了优秀的可视化能力和灵活的扩展能力。

　　平台特点：支持主流 GIS 厂商发布的标准 OGC 服务；支持简单高效、稳定灵活的插件扩展；支持 Chrome、FireFox、IE 等多种主流浏览器；支持多种地图和数据可视化效果。

2 KQGIS WebGIS 开发环境配置

2.1 KQGIS Server 平台安装

免安装版本,可直接解压,导入产品授权许可即可使用。注意:授权许可须命名为"KanqGIS.lic"(图 2-1)。

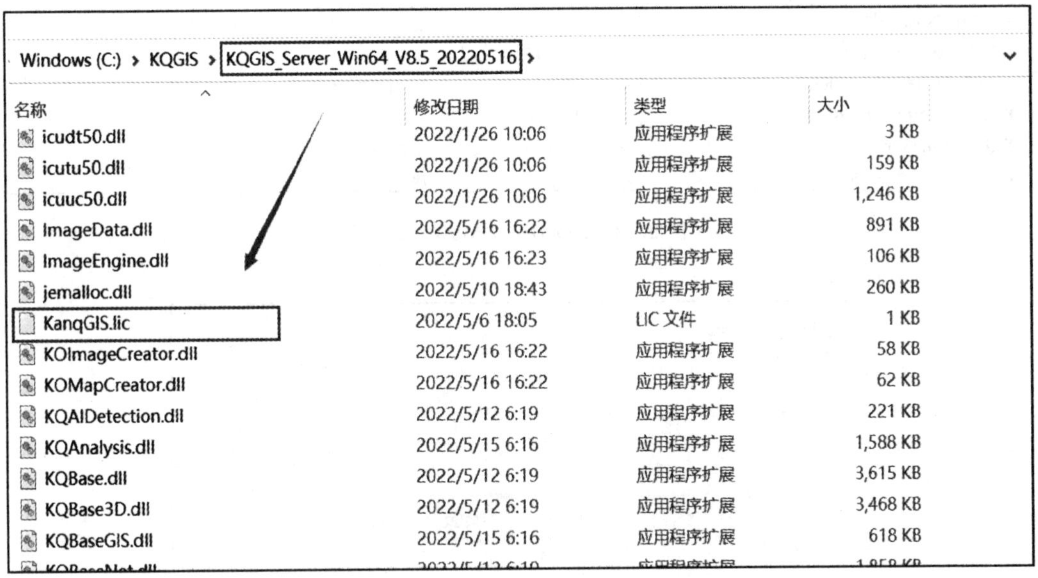

图 2-1 产品授权许可

以管理员身份打开"KQGISServer.exe",进入服务启动器界面,如图 2-2 所示。

第一步,点击"安装",安装服务启动器,如图 2-3 所示。

第二步,点击"启动",启动服务启动器,如图 2-4 所示。

第三步,以管理员身份打开"KQDesktopPro.exe",点击"目录视图"→"KQGIS 服务器",鼠标右键点击"刷新"服务器列表,如图 2-5 所示。

第四步,双击"目录视图"→"KQGIS 服务器"→"添加 KQGIS 服务器",如图 2-6 所示。

第五步,设置"服务器 IP",测试通过,即可添加,如图 2-7 所示。

第六步,双击"127.0.0.1:8699/kqgis",启动服务器(图 2-8)。启动成功后,图标发生变化。

第七步,在浏览器输入框中输入"127.0.0.1:8699"进入 KQGIS Server 管理平台。若出现发布的服务目录及子目录则表明配置成功。

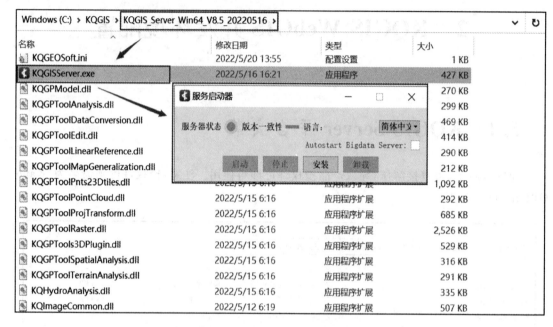

图 2-2 KQGIS Server 解压预览

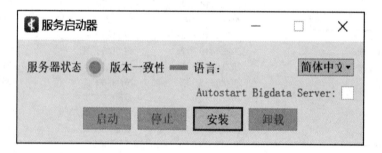

图 2-3 KQGIS Server 服务器启动 1

图 2-4 KQGIS Server 服务器启动 2

2　KQGIS WebGIS 开发环境配置

图 2-5　KQGIS Server 服务器添加 1

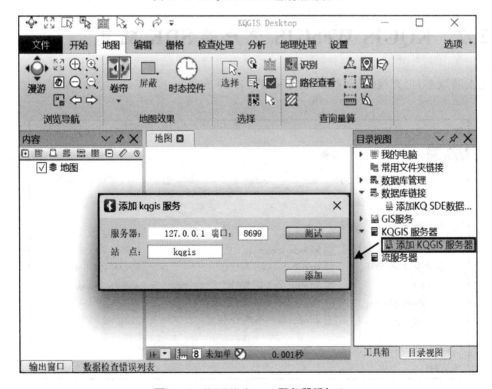

图 2-6　KQGIS Server 服务器添加 2

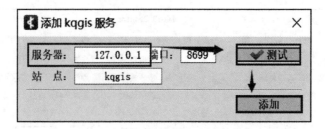

图 2-7　KQGIS Server 服务器添加 3

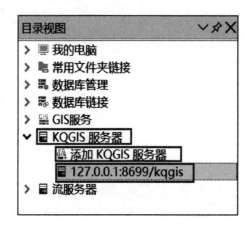

图 2-8　KQGIS Server 服务器添加 4

2.2　KQGIS WebGIS 客户端 SDK 配置

免安装版本,可直接解压,解压完成后如图 2-9 所示。

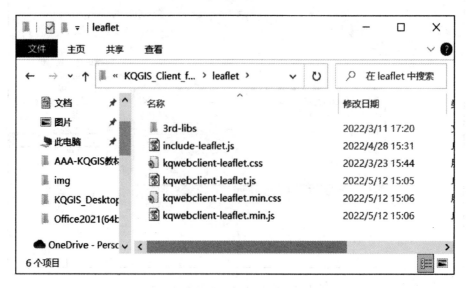

图 2-9　KQGIS Client for Leaflet 解压预览

将文件夹拷贝到工程根目录中获取文件后,只需要像普通的 JavaScript 库一样用<script>标签引入即可。首先新建一个 HTML 文件,在 head 标签中引入相关 JS 文件,再填入前面下载的依赖包解压后的文件地址,例如:

```
<!DOCTYPE html>
<html>
<head>
<meta charset="UTF-8">
<script>window.lib_path="../dist/leaflet/3rd_libs"</script>
<script type="text/javascript" src="../dist/leaflet/include-leaflet.js"></script>
<script type="text/javascript" src="../dist/leaflet/kq-release.js"></script>
</head>
</html>
```

即可完成对 SDK 的配置。

2.3 示例数据库配置

免安装版本,可直接解压,解压完成后如图 2-10 所示。

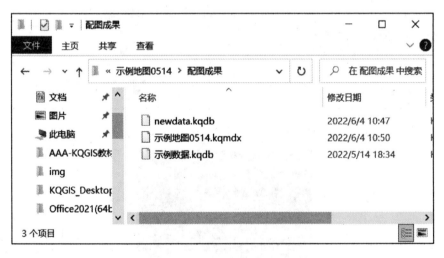

图 2-10 示例数据解压预览

第一步,双击"目录视图"→"数据库管理"→"数据库链接"→"添加 KQ SDE 数据库链接",并选中"KQSpatialDatabase"(图 2-11)。

第二步,点击右边的"文件夹"按钮,选择已经解压的 *.kqdb 文件(以"示例数据.kqdb"为例)(图 2-12)。

第三步,点击"保存",并在创建数据库界面中点击"确定",即可完成对示例数据库的配置(图 2-13)。

图 2-11 示例数据库配置 1

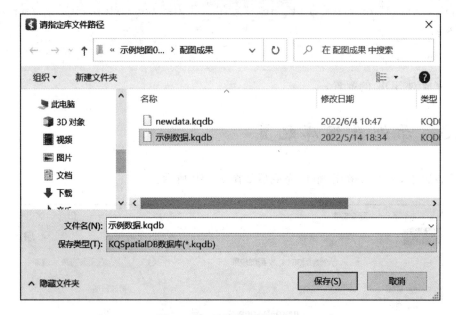

图 2-12 示例数据库配置 2

图 2-13 示例数据库视图

2.4 矢量地图服务发布

参考 2.1,在完成服务器的添加后,鼠标右键点击"KQGIS 服务器"→"127.0.0.1:8699"→"添加服务"按钮,弹出"添加 GIS 服务"对话框,输入服务名(服务名称只能输入数字、小写字母、下划线,并且只能为半角字符)、别名、服务类型(选择"苍穹二维地图服务")、描述,这里服务名以"basemap"为例,点击"下一步"按钮,如图 2-14 所示。

图 2-14 矢量地图服务发布 1

地图文件路径:点击"浏览"按钮,选择解压的"KQGIS 示例地图"文件夹中的"示例地图 0514.kqmdx"文件(图 2-15)。

图 2-15 矢量地图服务发布 2

瓦片存储路径:支持"本地目录"和"MongoDB"两种类型。

本地目录(推荐):当选择该类型时,可留空使用默认存储目录,也可指定存储目录。点击"瓦片目录"后面的"浏览"按钮,可选择要指定的目录(图 2-16)。

图 2-16 矢量地图服务发布 3

MongoDB:若需使用该类型,首先需要进行 MongoDB 安装和配置,配置完成后,选择"MongoDB"设置连接参数 MongoDB,设置好连接参数后,点击"测试连接"按钮,弹出"连接数据库成功",表示连接参数设置正确。同时也支持输入用户名、密码的连接方式。如图 2-17 所示。

图 2-17 矢量地图服务发布 4

点击"结束"按钮,发布服务,如图 2-18 所示。

其目录视图如图 2-19 所示。

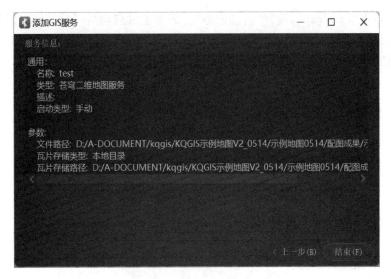

图 2-18 矢量地图服务发布 5

图 2-19 矢量地图服务目录视图

鼠标右键点击"basemap",选择"启动服务"。服务正常启动后,鼠标右键选择"浏览 REST"服务,系统会自动弹窗进入 REST 服务浏览界面。若 basemap 可以正常显示,则表示服务已发布成功(图 2-20)。

图 2-20 地图服务图像预览

鼠标右键点击"basemap",选择"启动服务"。服务正常启动后,鼠标右键选择"浏览示例"页面,系统会自动弹窗进入地图服务信息页。在这里可以看到地图服务的坐标、参考系、图层名、分辨率等信息(图2-21)。

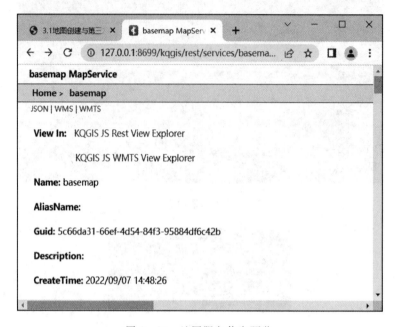

图2-21　地图服务信息预览

2.5　二维影像服务发布

二维影像服务发布前需要在数据库里新建镶嵌栅格数据集导入的数据,在参考2.3完成数据库配置后,数据库链接菜单如图2-22所示。

图2-22　数据库链接菜单

鼠标右键点击"KQSpatialDB",选择"新建"→"镶嵌栅格数据集"(图2-23)。

填写好数据集名称,坐标系选择"Geographic Coordinate Systems"→"Asia"→"Beijing 1954"。后两项留空即可(图2-24)。

填写完成后,点击"确定"发布。发布完成后效果如图2-25所示(若无上面两项,可能是数据库未连接,尝试双击连接数据库)。

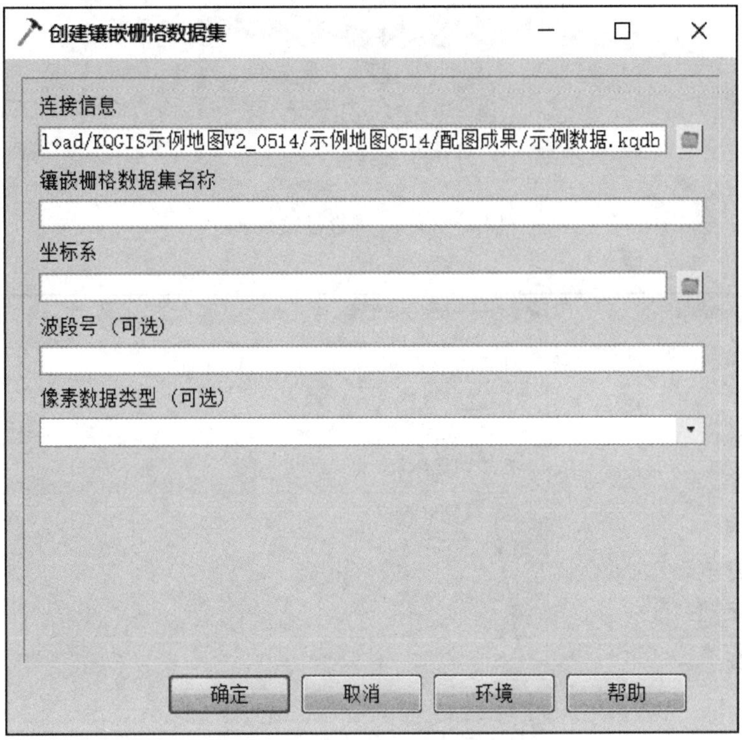

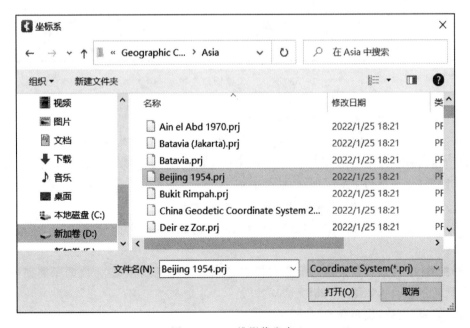

图 2-23　二维影像发布 1

图 2-24　二维影像发布 2

此时,鼠标右键点击"数据库",选择"为服务生成配置文件"→"影像服务",选择影像集并填写文件导出路径(图 2-26)。

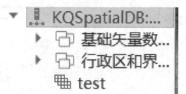

图 2-25 二维影像发布 3

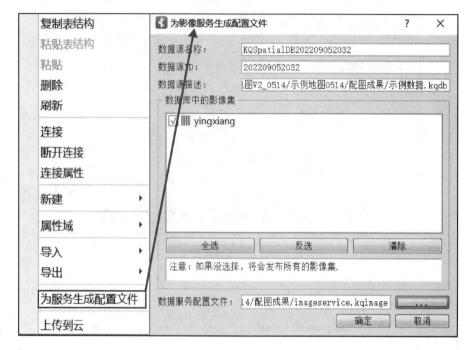

图 2-26 二维影像发布 4

点击"确定",系统提示配置文件导出成功。

参考 2.1,在完成服务器的添加后,右键点击"KQGIS 服务器"→"127.0.0.1:8699"→"添加服务"按钮,弹出"添加 GIS 服务"对话框,输入服务名(服务名称只能输入数字、小写字母、下划线,并且只能为半角字符)、别名、服务类型(选择"苍穹二维影像服务")、描述。这里服务名以"yingxiang"为例,点击"下一步"按钮,如图 2-27 所示。

点击"下一步"按钮,如图 2-28 所示,配置文件路径选择之前导出的 *.kqimage 文件(图 2-28)。

选择配置文件,也可以手动对配置文件进行修改,影像数据类型和数据库类型可以自定义(推荐默认不修改),如图 2-29 所示。

点击"结束"按钮,发布服务,如图 2-30 所示。

添加服务成功,如图 2-31 所示。

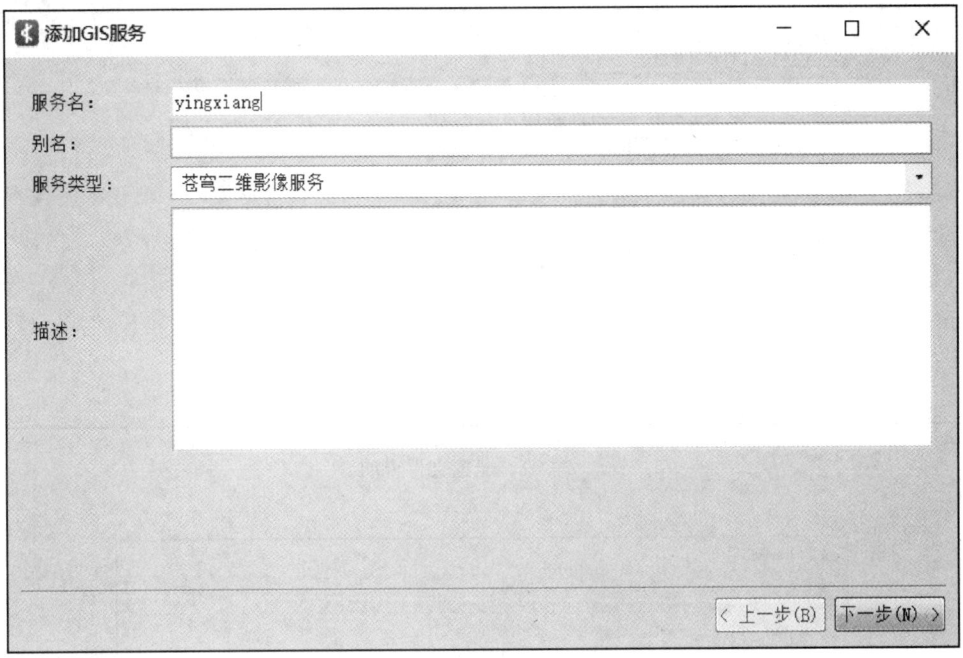

图 2-27　二维影像发布 5

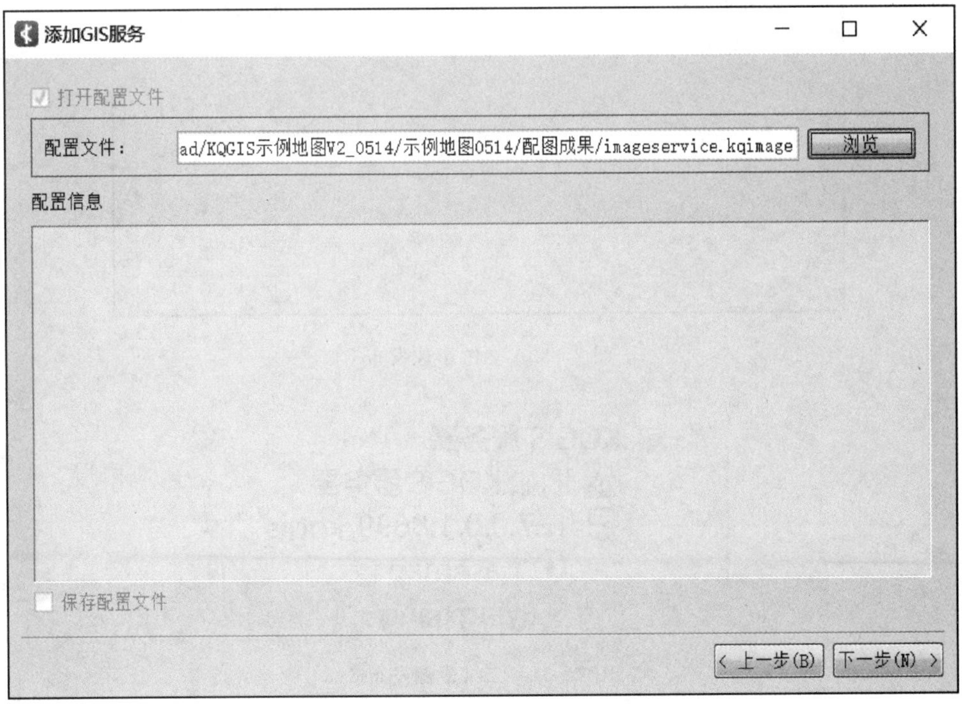

图 2-28　二维影像发布 6

```
1  {
2      "ImgData" 自定义
3      "datasourceID":"120",自定义
4      "description":"Img KQSpatialDB数据库",影像类型  数据库类型（以实际的数据类型和数据库为准）
5      "ConnectionInfo":{
6          "connecttype":"KQSpatialDB",数据库
7          "filepath":"G:/hsq/db/hsq.kqdb" 数据库所在路径
8      },
9      "collections":[
10     [
11         {
12             "collectionName":"hsq" 服务名
13             "crs":"EPSG:2362" 影像数据坐标系
14         }
15     ]
16  }
17  }
```

图 2 - 29 配置文件说明

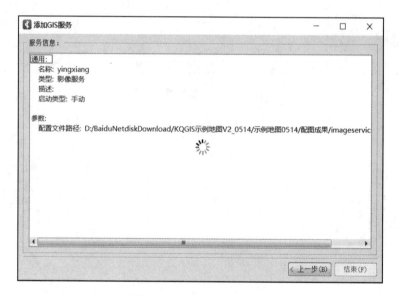

图 2 - 30 二维影像发布 7

图 2 - 31 二维影像发布 8

2.6 WMTS、WMS、WFS 服务发布

2.6.1 WMTS

WMTS 是 OGC 制定的一种发布瓦块地图的 Web 服务规范，WMS 主要是动态地图，WMTS 是地图服务器预先制作好的瓦块。在 GIS 领域，金字塔技术一直是一种基础性技术，WMTS 规范专门制定了针对切片请求的格式。利用这种技术，前端可以快速展示出指定级别的地图或影像。

如需添加 WMTS 服务：第一步，双击"目录视图"→"GIS 服务"→"WMTS 服务"；第二步，鼠标右键点击"WTMS 服务"，选择"新建地图连接"，填写名称地址等信息（以天地图地形图为例），点击"获取图层"，即可在右侧显示图层的比例尺、坐标等信息（图 2-32）；第三步，点

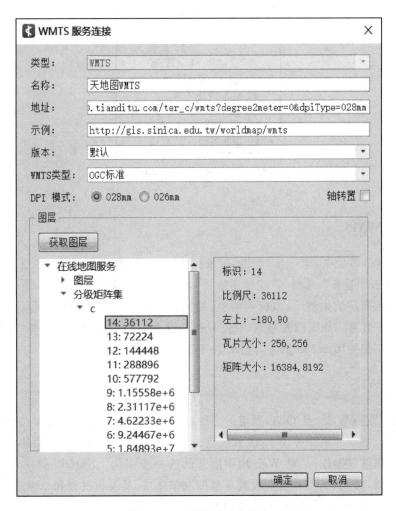

图 2-32 WMTS 服务连接

击"确定",即可在目录视图中查看已经添加的 WMTS 服务(图 2-33)。如需修改地址,只需鼠标右键点击"WMTS 服务",选择"属性"即可进入编辑界面。若不想使用 WMTS 服务,选择"断开连接"。

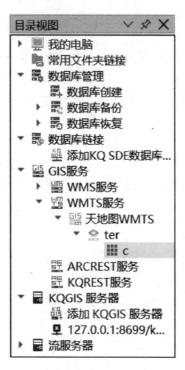

图 2-33　WMTS 目录视图

2.6.2　WMS

如需添加 WMS 服务:第一步,双击"目录视图"→"GIS 服务"→"WMS 服务";第二步,鼠标右键点击"WMS 服务",选择"新建地图连接",填写名称、地址等信息后,点击"获取图层",即可在右侧显示图层的比例尺、坐标等信息;第三步,点击"确定",即可在目录视图中查看已经添加的 WMS 服务。如需修改地址,只需鼠标右键点击"WMS 服务",选择"属性"即可进入编辑界面。若不想使用 WMS 服务,选择"断开连接"(图 2-34)。

2.6.3　WFS

Web 要素服务(WFS)是 OGC 制定的一种在互联网上对矢量地理要素及数据进行操作,包括检索、插入、更新和删除等 Web 服务规范。返回的是要素级的 GML 编码,并提供对要素的增加、修改、删除等操作,是对 Web 地图服务的进一步深化。KQGIS 提供对 WFS 服务的发布及查询功能,对 WFS 服务没有设置单独的发布接口,其发布过程和二维地图服务发布过程一致,参考 2.4 即可完成对 WFS 服务的发布。

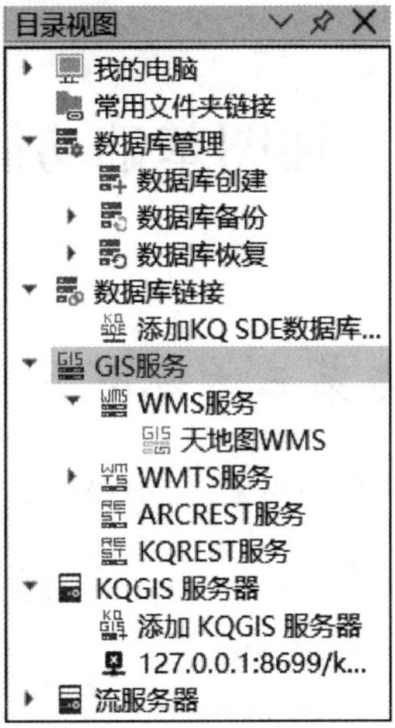

图 2-34 WMS 目录视图

3 地图基础功能

3.1 地图创建与第三方地图导入

参考 2.2 完成客户端 SDK 配置,并新建一个 HTML 页面,在页面中继续添加代码以创建地图,步骤如下。

第一步,在<head>标签中导入 SDK 文件以及网页样式文件。在<style>中设置"maptools"工具栏及其下拉列表的样式和尺寸信息。

```
<!DOCTYPE html>
<html lang='en'>
<head>
<meta charset='UTF-8'>
<script>window.lib_path='../../libs/kqwebclient/leaflet/3rd-libs';</script>
<script type='text/javascript' include="src="../../libs/kqwebclient/leaflet/include-leaflet.js"></script>
<script src='../../js/demo_setting.js'></script>
<link rel='stylesheet' href='../../css/demo.css' />
<link rel='stylesheet' href='../../libs/kendoui/styles/web/kendo.common.min.css' />
<link rel='stylesheet' href='../../libs/kendoui/styles/web/kendo.materialblack.min.css' />
<script src='../../libs/kendoui/js/kendo.ui.core.min.js'></script>
<style>
#maptools {
margin: 8px 7px;
position: absolute;
left: 0px;
top: 0px;
font-size: 12px;
z-index: 99999;
}
#maptools.k-dropdown {
```

```
width: 170px;
box-shadow: 2px 2px 2px #888888;
}
</style>
</head>
```

第二步，在＜body＞标签中绑定 onload 函数，并创建名为"map"的 div 容器以及"maptools"地图控件容器。

```
<body onload='onload()'>
<div id='map'></div>
<div id='maptools'>
</div>
```

以 2.4 中发布的中国地图数据服务为例，在＜script＞中添加代码，初始化地图信息。

第一步，设置 resolutions 分辨率数组，resolution 的实际含义是当前地图范围内 1 像素代表多少地图单位。这里的加粗部分需要读者参考地图服务信息页中的信息自行设置，具体查看方法参考 2.4。

例如，这里的 level0 是缩放层级为 0 时的地图分辨率，其值为 0.07605736072211919。后面每分一层，则将该值除以 2 的 n 次方（n 为 level ID），作为该级地图分辨率（图 3-1）。

Scales Info:

- **Height:** 256
- **Width:** 256
- **DPI:** 96
- **Levels of Detail:** (# Levels: 16)
    - **Level ID:** 0 dynamic
      Resolution:0.07605736072211919
      Scale:32000000
    - **Level ID:** 1 dynamic
      Resolution:0.038028680361059594
      Scale:16000000
    - **Level ID:** 2 dynamic
      Resolution:0.019014340180529797
      Scale:8000000
    - **Level ID:** 3 dynamic
      Resolution:0.009507170090264899
      Scale:4000000
    - **Level ID:** 4 dynamic
      Resolution:0.004753585045132449
      Scale:2000000

图 3-1　各层级分辨率信息

```
var res=[];
for (var i=0;i<18;i++) {
res.push(0.07605736072211919 / (Math.pow(2,i)));
}
```

第二步,初始化地图,调用 L.map 类实例化对象 map,设置地图加载的动画效果。

设定地图中心([34.3,100]代表东经100°,北纬34.3°),缩放层级7级,4326地图投影,KQClient for Leaflet通过投影类 L.Proj.CRS 可以方便地定义地图投影,支持设置范围、原点、比例尺数组以及分辨率数组。该类参数如表3-1所示。

为计算级别,options.scales/options.scaleDenominators/options.resolutions/options.bounds 必须指定一个,先后顺序已按优先级排列。当指定 options.bounds 时,第 0 级为一张 256 切片,包含整个 bounds,即 Math.max(bounds.getSize().x,bounds.getSize().y)/256。为保证切片行列号正确,options.origin/options.bounds 必须指定一个。当指定 options.bounds时,切片原点为 bounds 的左上角。

表 3-1 L.Proj.CRS 参数介绍

Name	Type	Default	参数定义
def	string		投影的 proj4 定义
origin	Array.<number>		可选,原点
scales	Array.<number>		可选,比例尺数组
scaleDenominators	Array.<number>		可选,比例尺分母数组
resolutions	Array.<number>		可选,分辨率数组
bounds	Array.<number>		可选,范围
dpi	number	96	可选,dpi

```
//初始化地图
map=L.map('map',{
//动画平滑效果
zoomtime:0.3,
customzoom:true,
center:[34.3,100],
zoom:7,crs:L.Proj.CRS('EPSG:4326',{
origin:[-180,90],resolutions:res,
bounds:L.bounds([-180,-180],[180,180])})})});
```

第三步,加载发布的地图,先创建 url 变量,其值为地图服务发布的本地 IP 地址。然后使用 L.KqGIS.tileMapLayer 类创建图层,该类具体参数如表3-2所示。

表 3-2 L.KqGIS.tileMapLayer 参数介绍

Name	Type	Default	参数定义
layers	string		图层名称
styles	Object		可选,图层样式
format	string	'image/png'	可选,图像格式
transparent	string	true	可选,服务返回的图像是否透明
version	string	'1.1.1'	可选,版本信息
zoomOffset	number	0	可选,比例尺的偏移值,默认情况下为0(表示不进行偏移)
minZoom	number	0	可选,最小比例尺
maxZoom	number	18	可选,最大比例尺
attribution	string		可选,版权信息

将 url 作为参数传入,填写其他区图层可选参数。这里挑选了前6个图层作为例子,若留空,则默认选择所有图层。

```
let url = 'http://127.0.0.1:8699/kqgis/rest/services/basemap/map';
L.KqGIS.tileMapLayer(url,{
layers:[1,2,3,4,5,6],}).addTo(map);}
```

运行效果如图 3-2 所示。

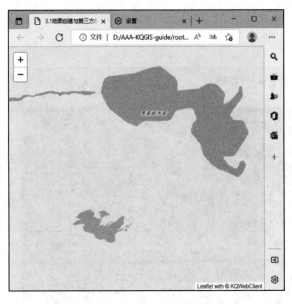

图 3-2 已发布地图加载效果预览

若想使用第三方地图,KQWebMap API for Leaflet 对多种互联网地图信息进行了封装,例如百度地图、天地图等。

以天地图为例,KQWebMap API for Leaflet 提供了 tiandituTileLayer 类,参数如表 3-3 所示。

表 3 - 3 tianditiTileLayer 参数介绍

Name	Type	Default	参数定义
layerType	string	'vec'	可选,图层类型(vec:矢量,cva:矢量注记,eva:英文矢量注记,img:影像,cia:中文影像注记,eia:英文影像注记,ter:地形,cta:地形注记)
minZoom	number	1	可选,最小比例尺
maxZoom	number	18	可选,最大比例尺
isDegree	number	false	可选,是否是经纬度
key	string	'defaultkey'	可选,map key
subdomains	Array.<number>	[0,1,2,3,4,5,6,7]	可选,子域名数组
className	string		可选,'tdtdark':深色地图,默认为原始效果

使用时只需使用 L. layerGroup 创建图层组,并用 addTo 方法,将图层组加入地图,随后用 tianditiTilelayer 加载对应图层(vec 矢量图层、cva 矢量注记图层等),用 addTo 方法将图层加入图层组,即可显示第三方地图。

```
layergroup=L. layerGroup();
layergroup. addTo(map);
L. KqGIS. tianditiTileLayer({ layerType: 'img',isDegree: true,zoomOffset: 1 }). addTo(layergroup);
```

运行效果如图 3 - 3 所示。

图 3 - 3 第三方地图加载效果预览

3.2 基本地图控件

L. control 是一个实现地图控件的基类,负责处理定位。所有其他的控件都是从这个类中延伸出来的,它包括一个参数值 position 用于决定空间位置位于地图的某一个角,可能的值是"topleft""topright""bottomleft"或"bottomright"。通过向地图添加控件的方式,实现对图层的放大、缩小、图层切换等交互操作,常用的控件如表 3-4 所示。

表 3-4　L. control 子类概览

控件	类名	参数定义
导航	L. control. pan	默认位于地图左上角,用于调整地图显示范围
鹰眼图	L. control. MiniMap	默认位于地图右下角,用于显示当前浏览区域的缩略图
图层切换	L. control. layers	默认位于地图右上角,用于切换矢量、影像、矢量注记等图层
卷帘	L. control. sideBySide	默认出现在地图中心,用于实现矢量影像图层的对比显示
比例尺	L. control. scale	默认位于地图左下角,根据缩放程度确定
定位	L. control. locate	默认位于地图左上角,用于快速返回当前定位位置
缩放	L. kqmap. control. zoomInButton、L. kqmap. control. zoomOutButton	位置由父控件决定,用于缩放地图

当添加控件时,首先初始化地图,然后通过 addTo 方法将控件添加到地图上。

3.2.1 导航控件

第一步,初始化地图,使用 L. map 类实例化 map 对象。设置地图中心、缩放等级。

```
function onload() {
var map=L. map("map",{center:[30.543,114.397],zoom:10,zoomControl:false,attributionControl:false});
```

第二步,使用 gaodeTileLayer(具体使用方法详见 5.5.4)导入 vec 矢量图层,并使用 addTo 方法将矢量图层加入 map 中作为底图。

```
L. kqmap. mapping. gaodeTileLayer({layerType:"vec"}). addTo(map);
```

第三步,使用 L. Control 类添加地图控件,pan 是地图导航控件,使用 addTo 方法将控件加入地图中,可直接在地图中查看控件效果。

```
L. control. pan(). addTo(map);}
```

运行效果如图 3-4 所示。

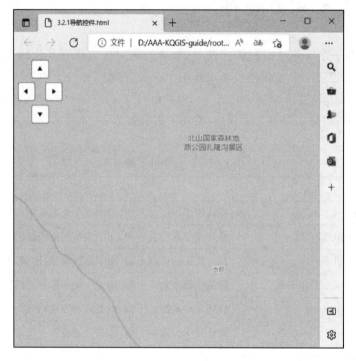

图 3-4 导航控件效果预览

3.2.2 缩放控件

参考上文中导航控件的添加,进行缩放控件的添加。操作步骤如下:第一步,初始化地图,设置地图中心及缩放等级;第二步,添加天地图矢量图层及矢量注记图层;第三步,添加控件,调整控件位置。

缩放控件添加方式与导航控件相似,因此不再对过程进行赘述,唯一不同之处在于添加地图控件时,子类名为"zoom",即调用缩放控件。

```
//添加地图控件
L.control.zoom().setPosition('bottomleft').addTo(map);
```

运行效果如图 3-5 所示。

3.2.3 比例尺

比例尺表示图上一条线段的长度与地面相应线段的实际长度之比。公式为:比例尺＝图上距离:实际距离。比例尺控件添加方式与导航控件相似,唯一不同之处在于添加地图控件时,子类名为"scale",即调用比例尺控件。

```
//添加比例尺控件
L.control.scale().setPosition('bottomright').addTo(map);
```

3 地图基础功能

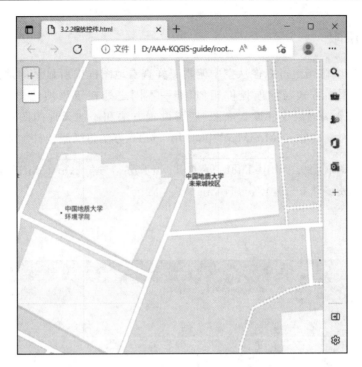

图 3-5 缩放控件效果预览

运行效果如图 3-6 所示。

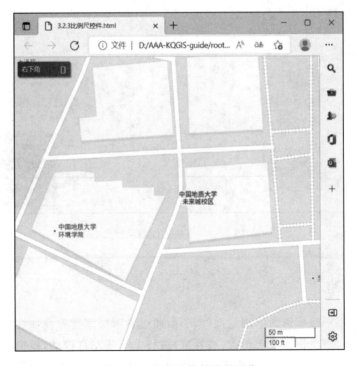

图 3-6 比例尺控件效果预览

3.2.4 鹰眼

顾名思义，在鹰眼图上可以像从空中俯视一样查看地图框中所显示的地图在整个图中的位置。鹰眼控件添加方式与导航控件相似，唯一不同之处在于添加地图控件时，子类名为"MiniMap"，且需要为鹰眼控件设置底图，底图设置方法和添加天地图图层一致。最后将创建好的天地图作为参数导入"MiniMap"中，使用 addTo 方法将鹰眼控件加入地图。

```
var img=L.KqGIS.tiandituTileLayer({ layerType："img",attribution：null });
new L.Control.MiniMap(img).addTo(map);
```

运行效果如图 3-7 所示。

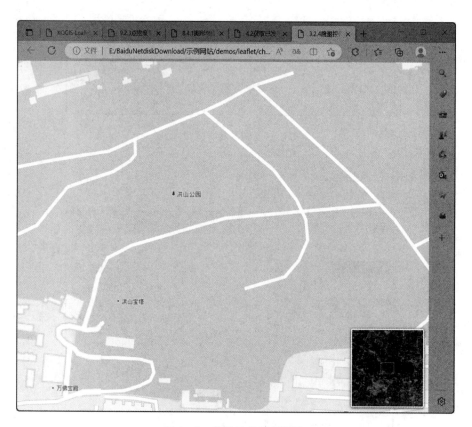

图 3-7 鹰眼控件效果预览

3.2.5 卷帘

地图卷帘常用于遥感影像的浏览操作，一般通过卷帘操作查看叠加在一起的两幅影像的差别，其操作效果为：鼠标在上层影像（被卷帘）进行上下或左右拉动，模拟一种将影像上下或左右卷起的动作，从而呈现出下层影像。与 3.2.4 中的鹰眼设置类似，卷帘也需要设置相应的底图。使用 L. kqmap. mapping. tiandituTileLayer 新建 vec 矢量图层及 img 影像图层,将其作为参数输入卷帘控件 L. control. sideBySide,使用 addTo 方法将其加入地图中。

```
//添加天地图矢量图层
var vec=L. kqmap. mapping. tiandituTileLayer({
layerType: "vec",
attribution: null
}). addTo(map);
//添加天地图影像图层
var img=L. kqmap. mapping. tiandituTileLayer({
layerType: "img",
attribution: null
}). addTo(map);
//添加天地图矢量注记图层
L. kqmap. mapping. tiandituTileLayer({
layerType: "cva",
attribution: null
}). addTo(map);
L. control. sideBySide(vec,img). addTo(map);
```

运行效果如图 3-8 所示。

图 3-8 卷帘控件效果预览

3.2.6 图层探查(滤镜)

图层探查是一个功能型的控件,当 map 加载多个图层的时候,有的图层被覆盖了,如果想看到图层就可以通过探查图层小插件,相当于将上面的图层裁剪一块,具体实现代码如下。由于 Leaflet 不直接提供图层探查控件,需要导入 magnifyingGlass 扩展包。首先用 gaodeTileLayer 创建 img 影像图层,使用 addTo 方法将影像图层加入地图中,再使用 gaodeTileLayer 实例化名为"gaode"的图层。新建 magnifyingGlass 变量,设置 zoomoffset 缩放大小为 1(原始大小)。然后设置滤镜图层为之前创建的"gaode"。完成之后,使用 L.control.magnifyingglass 创建滤镜控件并使用 addTo 方法将其加入地图。

```
L.KqGIS.gaodeTileLayer({ layerType: 'img' }).addTo(map);
var gaode=L.KqGIS.gaodeTileLayer({ layerType: 'vec' });var magnifyingGlass=L.magnifyingGlass({zoomOffset:1,layers:[gaode]});L.control.magnifyingglass(magnifyingGlass).addTo(map);
```

运行效果如图 3-9 所示。

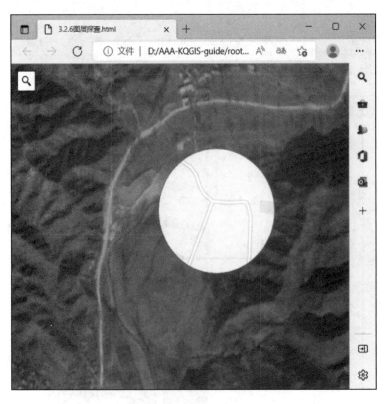

图 3-9 图层探查控件效果预览

3.3 地图图层切换

图层切换控件,可以实现多个图层切换功能,具体实现步骤如下:第一步,使用 L. kqmap. mapping. tiandituTileLayer()初始化多个图层;第二步,定义各图层名字,初始化地图,设置地图中心、缩放等级,将 vec 及 cva 作为默认图层加入地图;第三步,使用 L. control. layers 创建图层切换控件。其中,baseLayers 参数表示可以切换的图层,以单选形式展示;overlayers 表示覆盖图层,以 checkbox 形式展示;baseLayers 可以是由 L. tileLayer 创建的切片图层,或者是 L. layerGroup 创建的图层分组。

```
L. control. layers(baseMaps,overlays). addTo(map). setPosition("topleft");
```

运行效果如图 3-10 所示。

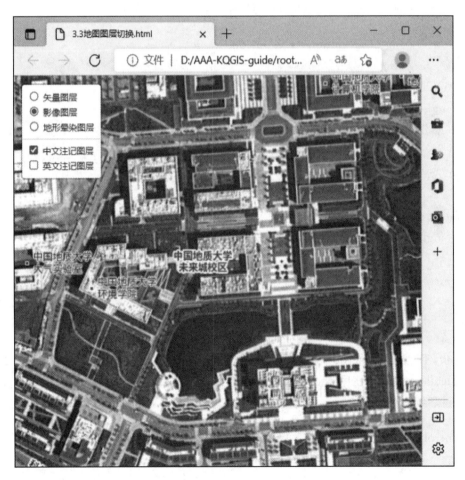

图 3-10 地图图层切换效果预览

3.4 地图数据测量

KQClient for Leaflet 支持距离量算和面积量算,实现方法均为调用 REST 地图服务接口进行查询。

3.4.1 距离量算

距离量算通常用于确定实际地形中两点或多点之间的连线总距离,具体实现方法如下。

第一步,初始化地图,设置地图中心和缩放级别。

第二步,使用 L.Control 类自定义 LengthCalculation 控件,使用 DomUtil 类的 create 方法创建 length_calculation 容器,并用 control_html 变量承接,新建 template 容器,在容器内创建距离量算后的结果视窗容器,设置其 id 为 length 以及"计算长度"按钮。使用 settimeout 方法,把 template 加入 length_calculation 容器中,返回 control_html。

```
L.Control.LengthCalculation=L.Control.extend({
onAdd: function(map) {
let control_html=L.DomUtil.create('div','length_calculation');
let template=`<div id='maptools'>
<input id='length'type='text' class='k-input k-textbox'>
<button id='calculate'>计算长度</button>
</div>
`;
setTimeout(function() {
 $('.length_calculation').append(template);
});
return control_html;
}
});
```

第三步,创建 lengthCalculation 函数,以 options 作为输入参数。调用该方法创建 LengthCalculation 容器,使用 addTo 方法把新建的容器添加到地图的左上角。options 参数设置方法见后续步骤。

```
L.control.lengthCalculation=function(opts) {
return new L.Control.LengthCalculation(opts);
};
L.control.lengthCalculation({ position: 'topleft' }).addTo(map);
```

第四步,新建高德地图矢量图层。使用 L. polyline 创建折线,确定每个点的坐标。设置线的颜色为红色并添加入地图。使用 map. fitBounds 调整地图显示范围为折线范围。

```
L. KqGIS. gaodeTileLayer({ layerType: 'vec' }). addTo(map);
let polyline=L. polyline([
[30.490096,114.561665],
[30.480629,114.573338],
[30.499563,114.585011]
],{ color: 'red' }). addTo(map);
map. fitBounds(polyline. getBounds());
```

第五步,使用 setTimeout 方法自动调用函数,初始化 kendo 容器,将计算按钮、计算结果分别绑定为 kendobutton 和 kendonotification。完成绑定后,使用 disableScrollPropagation 终止事件派发,阻止第二步中设置的容器被分派到其他 Document 节点。

第六步,设置服务链接 serviceurl,并将"calculate"按钮绑定鼠标单击事件,用户点击按钮时,使用 KqGIS. ServiceSRS 新建 geoSRS 坐标系,值为 EPSG4326,outSRS 坐标系,值为 EPSG3857;使用 KqGIS. MeasureParams 类实例化对象 params 以储存相应参数。

```
let serviceUrl='http://127.0.0.1:8699/kqgis/rest/services/basemap/map';
//计算长度
$('#calculate'). on('click',function() {
let geoSRS=new KqGIS. ServiceSRS({
type: KqGIS. ProjectSystemType. EPSG,
value: '4326'
});
let outSRS=new KqGIS. ServiceSRS({
type: KqGIS. ProjectSystemType. EPSG,
value: '3857'
});
let params=new KqGIS. MeasureParams({
data: polyline,
geoSRS: geoSRS,
outSRS: outSRS
});
```

第七步,调用 L. KqGIS. geometryAnalysisService 的子类 distance 距离量算服务,将 params 和 serviceurl 作为参数输入。该类的参数如表 3-5 所示。

表 3-5 距离量算 L. KqGIS. geometryAnalysisService 参数介绍

Name	Type	参数定义
url	String	服务器的 url 地址
data	GeoJSONObject	要计算面积或长度的图形对象，必选项。类型为 GeometryCollection 或 Polygon 等 GeoJSON 对象
geoSRS	String	图形的坐标所属的空间参考，必选项
outSRS	String	计算距离所使用的空间参考，必选项

若成功，则将返回的 result 在结果视窗中显示，并使用 flyToBounds 将地图显示范围调整至多段线的范围，提示"请求成功"。否则，提示"请求失败"。

```
    L. KqGIS. geometryAnalysisService(serviceUrl). distance(params, onSuccess, onFailed);
    });
    let onSuccess=function(response) {
    let result=response. result;
    $('#length'). val(result. result[0]. distance + '米');
    map. flyToBounds(polyline. getBounds());
    notification. show({
    message：'请求成功'
    },'info');
    };
    let onFailed=function() {
    notification. show({
    message：'请求失败'
    },'info');
    };
    });
    }
```

运行效果如图 3-11 所示。

3.4.2 面积测算

面积测算通常用于确定实际地形中几个点连成的多边形的总面积。代码部分与距离量测一致，唯一不同的是需要调用 L. KqGIS. geometryAnalysisService 的子类 distance 距离量算服务，将 params 和 serviceurl 作为参数输入。该类的参数如表 3-6 所示。

3 地图基础功能

图 3-11 距离量算效果预览

表 3-6 面积测算 L. KqGIS.geometryAnalysisService 参数介绍

Name	Type	参数定义
url	String	服务器的 url 地址
data	GeoJSONObject	要计算面积或长度的图形对象,必选项。类型为 GeometryCollection 或 Polygon 等 GeoJSON 对象
geoSRS	String	图形的坐标所属的空间参考,必选项
outSRS	String	计算距离所使用的空间参考,必选项

若成功,则将返回的 result 在结果视窗中显示,并使用 flyToBounds 将地图显示范围调整至矩形框的范围,提示"请求成功"(图 3-12)。否则,提示"请求失败"。

3.5 地图信息标注

在 WebGIS 生产环境中,通常需要对地图进行标注以完成对地点的介绍说明,或是对点位的突出强调,Kqclient for Leaflet 提供丰富的标注功能,如需调用,需进行如下操作:第一步,初始化地图,设置中心点及缩放等级;第二步,初始化标注点阵数组(文字/图标/图文);第三步,使用 addTo 方法将点阵添加入图层。下面将详细介绍标注过程。

图 3-12 面积测算效果预览

3.5.1 图片标注

使用预先设置好的图片来标注点位,易于识别但缺少对点位的描述。具体实现步骤如下。

第一步,初始化地图,设置地图中心和缩放等级。

第二步,初始化标注图标 icon,设置图标 url 及大小,初始化标注点阵数组,使用 for 循环遍历 hotels.js 中 hotels.length 的值,并将创建一个大小为该值的数组,命名为"hotel"。

```
var icon=L.icon({
iconUrl:'../../images/hotel.png',
iconSize:[24,24],// size of the icon
});
for (let i=0; i < hotels.length; ++i) {
let hotel=hotels[i];
```

第三步,使用 L.marker 类实例化一个对象,以 hotel.lat,hotel.lon 酒店经纬度作为参数输入,默认图标设置为 icon。对每个图标使用 bindtoollip 方法添加图层显示文本,内容为第"+(i+1)+"个旅馆,并将其位置设定为图层上方,偏移量为(0,-1)。最后使用 gaodeTileLayer 添加矢量图层 vec,并使用 addTo 方法将矢量图层加入地图。

```
L.marker([hotel.lat,hotel.lon],{icon:icon}).addTo(map)
  .bindTooltip("<span style='font-size:14px;'>第"+(i+1)+"个旅馆</span>",
{direction:"top",offset:L.point(0,-1)})
}
L.KqGIS.gaodeTileLayer({layerGroup:"vec"}).addTo(map);
}
```

运行效果如图 3-13 所示。

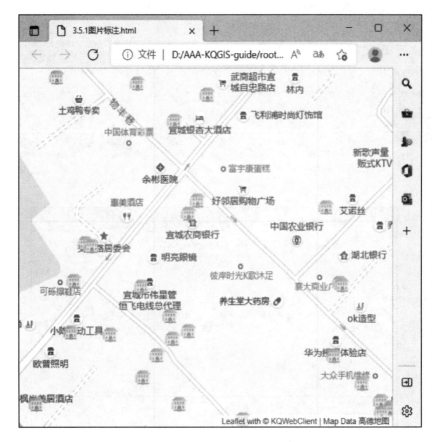

图 3-13 图片标注效果预览

3.5.2 文字标注

使用文字对目标点位进行描述,能够显示地图视野以外的信息,但不易识别。实现方法与 3.5.1 中的基本一致,故将核心代码展示如下。

此处直接使用 L.divIcon 创建文字点的 icon,该类使用 div 要素而非图片来轻量级地显示注记的图标。该类参数包括①iconSize:图标的像素大小,也可以通过 CSS 设置。②class-

Name：用于对图标自定义的类名，默认为 leaflet-div-icon。③html：在 div 要素中自定义的 HTML 代码，默认为空。④iconAnchor：图标提示的坐标（在左上角），图标是对齐的，所以这个点是注记的地理位置。如果大小是指定的，则位于中心处，也可以在 CSS 中设置负边界。

```
//文字点
for (let i=0; i < fontMarkers.length; ++i) {
    var font=fontMarkers[i];
    var icon=L.divIcon({
        html：'第' + (i + 1) + '个文字点',
        className：'my-div-icon',
        iconSize：75
    });
    L.marker([font.lat,font.lon],{icon：icon}).addTo(map)
    .bindTooltip("<span style='font-size：14px;'>第" + (i + 1) + "个文字点</span>",{direction："top",offset：L.point(0,-30)})
}
```

运行效果如图 3-14 所示。

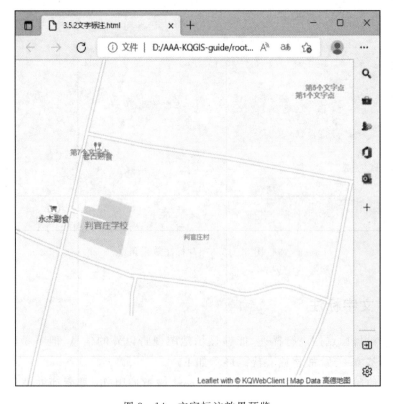

图 3-14　文字标注效果预览

3.5.3 Popup 标注

同时使用文字和图片对标注点位进行描述,兼具二者的优点,但在点位过多时会产生遮挡,影响看图,故在图文标注的基础上进行改进,对每个点位构建交互实体,点击"交互"后,弹出当前点位的图文信息,从而避免遮挡。其实现方法与 3.5.1 中的基本一致,故将核心代码展示如下。

同样地,使用 divicon 类创建 icon,但不同的是,在使用 L.marker 类创建图标点时,使用.bindPopup方法,为每个点绑定 popup 视窗。

```
features.map(function (feature,i) {
var tips="第" + (i + 1) + "张照片";
var icon=L.divIcon({
className: className,
iconSize: iconSize,
html: '<div title="' + tips + '" style="background-image:url(./../../images/thumb.jpg)";></div>',
});
L.marker([feature.lat,feature.lon],{icon:icon}).addTo(map)
.bindPopup('<div style="width:250px;text-align:center;"><h4>' + tips + '</h4><img src="../images/landscapes.jpg"></div>'));
}
```

运行效果如图 3-15 所示。

3.5.4 聚合标注

该方法用于大量标注点的环境中,在缩放等级较高时,将若干个距离相近的点位融合为新的点位,在缩放等级较低时,再将融合的点位还原,避免幅面混乱影响看图。具体实现步骤如下。

第一步,初始化地图,设置地图中心和缩放层级。使用 gaodeTileLayer 添加高德地图 vec 矢量图层作为底图。

第二步,设置 icon 大小。新建类名"leaflet-marker-photo",该类样式已在<style>标签中设置。使用 L.markerClusterGroup 创建图层组对象 resultLayer。showCoverageOnHover:当你的鼠标经过一个聚合点时,会显示它包含的标记围成的范围。该类参数包括①zoomToBoundsOnClick:当你点击一个聚合点时,会快捷缩放至一个其对应范围适合显示的 zoom 级别;②spiderfyOnMaxZoom:当你在最大缩放级别点击一个聚合点时,我们会将其蜘蛛化,以便用户能看到它包含的所有标记(标注:如果聚合点内的所有内容在最大的缩放级别或在 disableClusteringAtZoom 选项指定的缩放级别上仍有聚合点,则在当前缩放级别上进行蜘蛛化);③removeOutsideVisibleBounds:如果聚合点及标记处于用户的显示区域

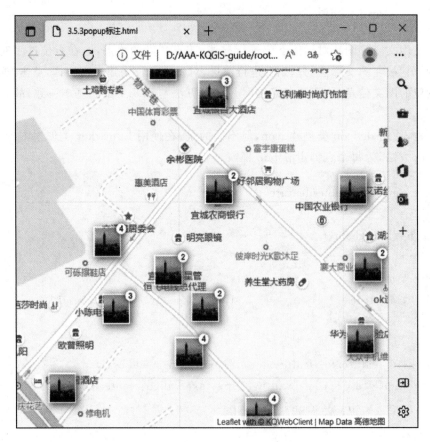

图 3 – 15 popup 标注效果预览

外,则出于性能考虑将其从地图上移除;④spiderLegPolylineOptions:允许给蜘蛛脚定义一个多边形选项 PolylineOptions,默认情况下是 { weight: 1.5, color: '#222', opacity: 0.5 }。⑤为方便预览,禁用 spiderfyOnMaxZoom、zoomToBoundsOnClick、showCoverageOnHover。

```
var iconSize=[40,40];
var className="leaflet-marker-photo";
var resultLayer=L.markerClusterGroup({
spiderfyOnMaxZoom: false,
//设置为 true 时显示聚类所占据的范围
showCoverageOnHover: false,
//设置为 true 时会向低一级聚类缩放
zoomToBoundsOnClick: false,
```

第三步,调用 features.map 方法创建要素,设置经纬度坐标 latlng,其值为 feature 的几何中心点,并转化为 EPSG3857 投影。最后将 latLng 作为参数输入到 L.marker 中创建标识并将其添加到 resultLayer 中,使用 addTo 方法将 resultLayer 加入地图中。

```
result.features.map(function (feature) {
var latLng = L.CRS.EPSG3857.unproject(L.point(feature.geometry.center));
resultLayer.addLayer(L.marker(latLng));
});
resultLayer.addTo(map);}
```

运行效果如图 3-16 所示。

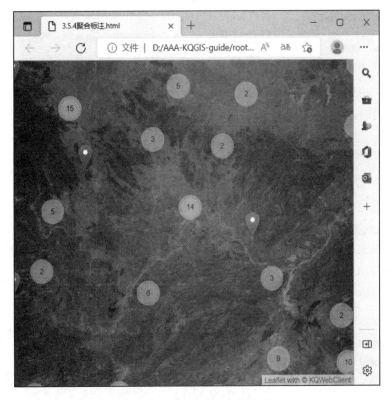

图 3-16 聚合标注效果预览

3.6 地图动画展示

为了更好地显示 WebGIS 系统的动态演示功能，通常需要使用要素动画来对一个地理信息过程进行演示，KQclient for Leaflet 提供丰富的标注功能，如需调用，需进行如下操作：第一步，初始化地图，设置地图中心点并调整至合适的缩放等级；第二步，初始化点线面要素数据；第三步，设置动画参数，包括时间间隔、速度、移动范围等；第四步，为动画绑定鼠标事件，当识别到鼠标输入事件时触发。

3.6.1 要素动画

它通常为针对要素本身的动态效果演示，包括拉伸、旋转、弹跳等。以要素弹跳动画为例，具体实现步骤如下。

第一步，初始化地图，设置地图中心和缩放等级，使用 osmTileLayer 加载 osm normal 图层，并使用 addTo 方法将图层加入地图。

第二步，新建变量 bounds 作为动画演示范围，使用 L. Marker. setBouncingOptions 设置弹跳的高度及速度。设置弹跳动画的图标点，个数为 20。随机生成其经纬度坐标，以 bounds 数组的值作为取值范围。

```
var bounds=[[30.343,114.197],[30.743,114.597]];
L. Marker.setBouncingOptions({
bounceHeight：40,
bounceSpeed：60
});
_.times(20,function () {
var lat=_.random(bounds[0][0],bounds[1][0]);
var lng=_.random(bounds[0][1],bounds[1][1]);
```

第三步，以经纬度坐标为参数，使用 L. marker 类生成图标点，添加跳跃动画选项方法，将跳跃高度 bounceHeight 设置为 20。添加 bounce 方法启动跳跃动画，添加鼠标事件，点击以后对 this 指针指向的对象（鼠标所选目标）启动 toggleBouncing 方法，即鼠标点击"开始跳跃"动画，随后将 marker 点添加入地图。

```
L. marker([lat,lng])
. setBouncingOptions({
bounceHeight：20
})
. bounce()
. on('click',function () {
this. toggleBouncing();
}). addTo(map);
});
```

第四步，添加一个鼠标点击事件，使用 L. Marker. stopAllBouncingMarkers 来停止所有动画事件。

```
map.on('click',function() {
L. Marker. stopAllBouncingMarkers();
});
}
```

运行效果如图 3-17 所示。

图 3-17 要素动画效果预览

3.6.2 要素移动

要素移动通常为要素平移或沿固定路线进行移动，其核心代码如下。

第一步，初始化地图，设置地图中心及缩放等级，使用 gaodeTileLayer 添加 vec 矢量图层，并使用 addTo 方法将图层加入地图作为底图。

第二步，设置经纬度坐标范围 latlng1，以 latlng1 为参数，使用 L.polyline 生成线，设置线颜色为黄绿色，将其加入地图。使用 L.icon 生成点，填写点 url、大小、角度等信息。

```
var latlng1=[[29.65,91.13],[30.67,104.07],[23.13,113.27],[26.08,119.30]];
//初始化线
L.polyline(latlng1,{ color: 'yellowgreen' }).addTo(map);
//点图标
var Icon=L.icon({
iconUrl: '../../images/bike.png',
iconSize: [25,40],
iconAnchor: [12.5,40],
});
```

第三步，使用 setTimeout 方法调用轨迹函数，先初始化图标点，设置其位置并添加入地图，使用 marker 类的 slideTo 方法，设置点运动的终点，将动画时间设定为 1000ms。然后调用 map.removeLayer 方法清除 marker1_1 图层，设定其起始时间为 0ms，终止时间为 1500ms。以相同方法设定第二段、第三段轨迹。

```
//运动轨迹
//第一段
setTimeout(function () {
var marker1_1=L.marker([29.65,91.13],{icon:Icon}).addTo(map);
marker1_1.slideTo([30.67,104.07],{duration:1000});
setTimeout(function () {
map.removeLayer(marker1_1);
},1500);},0);
//第二段
setTimeout(function () {var marker1_2=L.marker([30.67,104.07],{icon:Icon}).addTo(map);
marker1_2.slideTo([23.13,113.27],{duration:2000});
setTimeout(function () {
map.removeLayer(marker1_2);},2500);},1500);
//第三段
setTimeout(function () {
var marker1_3=L.marker([23.13,113.27],{icon:Icon}).addTo(map);
marker1_3.slideTo([26.08,119.30],{duration:1000});
},4000);
```

运行效果如图3-18所示。

图3-18 要素移动效果预览

3.6.3 动态航线

动态航线是指用于描绘物体运动的过程及实时性的航线。实现动态航线的原理很简单，就是每隔一段时间就更新线段的位置，连贯起来就是线段流动的效果。其核心代码如下。

第一步，初始化点图标以及点运动轨迹数组 mockData。

```
//点图标
var Icon=L.icon({
iconUrl:'../images/bike.png',
iconSize：[25,40],
iconAnchor：[12.5,40],
});
// 模拟数据
var mockData=[
[39.904,116.408],[36.663,117.009],[32.048,118.769],[31.213,121.445],
[30.319,120.165],[26.071,119.303],[23.108,113.265],[25.051,102.702],
[30.67,104.071],[36.068,103.751],[34.285,108.969],[30.573,114.279]
];
```

第二步，设置指针 index，间隔时间 interval 值为 1000ms。创建空数组 realtimeLine，使用 L.kqmap.realtimePolyline 类实例化生成线要素对象 realtimePolyline。

```
//模拟指针
var index=0;
//间隔时间(传的 interval 要一样)
var interval=1000;
//实时添加的全程线数据(初始一定为空数组)
var realtimeLine=[];
var realtimePolyline =L.kqmap.realtimePolyline();
// init()
//参数：source(func)
//参数：options(obj) -- interval 间隔时间
```

第三步，设置对象的初始化函数，创建变量 source 存储 mockData 中的值。调用对象的 draw 方法，将 source 和之前创建的空数组作为参数传入。设置绘制颜色为黄绿色，绘制时间为预设的间隔时间 interval，遍历整个 mockData 数组直到最后一位，期间重复绘制。

```
realtimePolyline.init(function () {
//模拟点数据
let source=mockData[index];
```

```
//动态画线 draw() options 中为可传参数,其他参数为必传参数
//参数:source(arr)实时得到的点数据
//参数:realtimeLine(arr)全程线数据
//参数:options(obj)color 画线颜色,icon(运动过程中 icon),interval
//参数:map
realtimePolyline.draw(source,realtimeLine,{
color:'yellowgreen',
icon:Icon,
interval
},map);
index++;
if (index === mockData.length) {
let timer=this._timer;
realtimePolyline.stop(timer,realtimeLine,{
icon:Icon,
interval,
},map);
}
},{
interval
});
```

运行效果如图 3-19 所示。

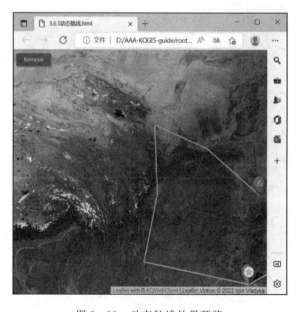

图 3-19 动态航线效果预览

4 KQGIS 目录服务

4.1 获取数据源列表及数据目录

数据源即 GIS 数据来源,通常可分为以下几种类型。①地图:各种类型的地图是 GIS 最主要的数据源,因为地图是地理数据的传统描述形式,我国的 GIS 系统图形数据大部分都来自地图;②遥感影像数据:遥感影像是 GIS 中一个极其重要的信息源,通过遥感影像可以快速、准确地获得大面积的、综合的各种专题信息,航天遥感影像还可以取得周期性的资料,这些都为 GIS 提供了丰富的信息;③数字数据:目前,随着各种专题图件的制作和各种 GIS 系统的建立,直接获取数字图形数据和属性数据的可能性越来越大,数字数据也成为 GIS 信息源不可缺少的一部分。

4.1.1 获取数据源列表

KQGIS 提供了一套数据源组织、存储的方法。以获取数据源列表为例,获取过程大致可分为以下几步。

第一步,初始化地图,添加高德地图矢量图层作为底图,初始化长宽并计算 offset。

```
//计算 offset
var clientWidth=document.body.clientWidth;
var clientHeight=document.body.clientHeight;
var width=644;
var height=260;
var offset={top:(clientHeight - height) / 2,left:(clientWidth - width) / 2};
```

第二步,datasource 容器绑定"kendobutton"按钮,并添加 async(异步)函数,创建 notification 变量绑定"kendoNotification"提示窗,设置其样式及内容。

```
$("#datasource - list").kendoButton({
click: async function () {
var notification= $("#notification").kendoNotification({
position: {
pinned: true,
```

```
bottom: 12,
right: 12
},
autoHideAfter: 2000,
stacking: "up",
templates: [{
type: "info",
template: <div>#= message #</div>
}]
}).data("kendoNotification");
```

第三步,请求数据,填写需要查询数据源的对象。设置查询参数 options 为 GetDatasourceList,对应 rest 接口→公共数据服务→数据源列表。使用 getParamString 解析参数并将结果及服务地址 url 作为参数输入 httpPostAsync 函数中。新建 response 变量承接函数结果(图 4-1)。

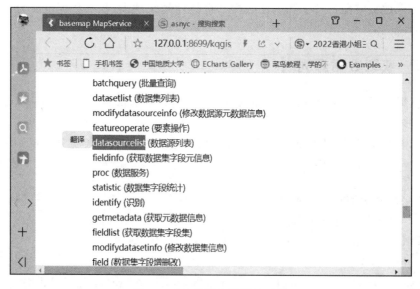

图 4-1　地图服务数据源列表

```
//请求数据
var url ='http://127.0.0.1:8699/kqgis/rest/services/basemap/map';
var options={
request: "GetDatasourceList"
};
let response=JSON.parse(await KQCommon.NetworkTools.httpPostAsync(url,KQCommon.NetworkTools.getParamString(options)));
```

第四步，若 response 的结果为 success，则调用回调函数，创建 data 数组，创建 datasourceList 数组承接 response 的结果，对 datasourcelist 执行 for 循环遍历每个 datasource，如果文件类型为 21(本地数据源)，则将其路径名作为 name 传入数组。否则，将其数据库名作为 name 传入。最后使用 push 方法将 datasource 的 id、type、name 等信息导入 data 数组中。

```javascript
if (response.resultcode == "success") {
notification.show({message："获取数据源列表成功。"},"info");
let data=[];
var datasourceList=response.result;
for (let i=0; i < datasourceList.length; ++i) {
let datasource=datasourceList[i];
let name="";
if (datasource.type == 21) {
name=datasource.path;
} else {
name=datasource.database;
}
data.push({
datasource_id: datasource.id,
typename: datasource.typename,
name: name,
})
}
```

第五步，使用 layui 提供的表格方法，添加弹窗及其标题、样式。随后把 data 数组中的数据导出，分为数据源 id、类型、名称 3 栏。若查询失败，则提示并返回。

```javascript
layer.open({
type:1,
title:"数据源列表",
content: '<table id="datasource-info"></table>',
offset: [offset.top + 'px',offset.left + 'px'],
skin: "layer_bg",
resize: false,
});
//填充数据表格
layui.table.render({
elem: '#datasource-info',
width: width,
```

```
height: height,
cols: [[ //标题栏
{field: 'datasource_id',title: '数据源 ID',width: 120},
{field: 'typename',title: '类型',width: 120},
{field: 'name',title: '名称',width: 400},
]],
data: data,
});
} else {
notification.show({message: "获取数据源列表失败。"},"info");
}}
}).data("kendoButton");}
```

运行结果如图 4-2 所示。

图 4-2　获取数据源列表效果预览

4.1.2　获取数据集目录

获取某地图服务的数据源后,如需继续查询当前服务的数据目录,只需修改 4.1.1 中第三步的查询参数,将 request 值改为 Getcatalog。如果想查询某个数据源下的数据目录,还需指定数据源 id。

```
//请求数据
var url='http://127.0.0.1:8699/kqgis/rest/services/basemap/map';
var options={request:"GetCatalog",datasourceId:selector.value()};
let response= JSON.parse(await KQCommon.NetworkTools.httpPostAsync(url,KQ-
Common.NetworkTools.getParamString(options)));
```

运行结果如图 4-3 所示。

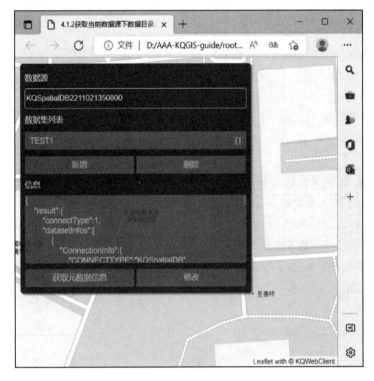

图 4-3　获取当前数据源下数据集目录

4.2　获取已发布的地图服务信息

在前文的介绍中,已经完成了对地图服务的发布,并且能够进入桌面端提供的 rest 示例页面中获取服务的各类信息,如需使用 rest 接口在 GIS 系统中对地图服务各类信息进行获取调用,需进行如下步骤。

4.2.1　地图服务图例信息

以地图服务图例信息为例,具体操作如下。

第一步,初始化地图,添加高德地图矢量图层。创建 makeCommand 对象承接 Make-

Command()函数,对_lists执行for循环,遍历其中所有元素,如果其中元素类型为function,则将其加入makeCommand对象中。调用makecommand对象的render属性与render()渲染方法(元素类型可能是function或其他实体对象)。调用show函数展示kqmapinfo。初始化图例图层并添加选项数组,包括名称和值。

```
//初始化渲染
var makeCommand=MakeCommand()
for (var i in _lists) {
if (typeof _lists[i] ! == "function") {
makeCommand. add(_lists[i])}}
makeCommand. render && makeCommand. render()
//初始化执行
_lists['kanqmapinfo']. show();
//初始化图例图层
_lists. legendlist()
var data=[
{ text: "kanq服务元数据信息",value: "kanqmapinfo" },
{ text: "WMS服务元数据信息",value: "wmsmapinfo" },
{ text: "WMTS服务元数据信息",value: "wmtsmapinfo" },
{ text: "WPS服务元数据信息",value: "wpsmapinfo" },
{ text: "地图整图",value: "entireimage" },
{ text: "图层信息",value: "layerinfo" },
];
var MakeCommand=function () {
return {
stacks: [],
add: function (command) {
this. stacks. push(command)
},
render: function () {
for (var i=0,command; command=this. stacks[i++];) {
command. render()
}
}
```

第二步,为selecter_service容器绑定"kendoDropDownList"下拉列表。为列表添加功能,当选项变化时,令value等于当前选项卡的值,然后调用show属性和show()方法展示当前值。

```
$("#selecter_service").kendoDropDownList({
dataTextField: "text",
dataValueField: "value",
dataSource: data,
index: 0,
change: function () {
var value = $("#selecter_service").val();
_lists[value].show && _lists[value].show()
}
});
}
```

第三步，使用 render 函数渲染 _list 列表，设置其样式为 html，将 body 添加到 html 中。

添加公共函数对象，当该函数被执行时，传入当前对象 id，将图例列表隐藏，进度条显示，图层组清空。然后展示当前 id 对应的对象。

添加 legendlist 函数，用 maplayerservice 承接地图服务 url。设置 params 参数，内容为需要展示的图层、格式、高度及宽度。调用 getLegend 函数，传入 params。用 results 接收 response 的结果。设置图例列表的容器 box 及其样式，递归地将子目录排列在父级目录下。使用 append 方法把 box 容器加入 body 中。

```
var _lists = {
render: function (id) {
var html = _html[id];
$('body').append(html)
},
commonExecute: function (id) {
$('#legendlist').hide()
$("#loading").show();
if (layergroup) {
layergroup.clearLayers();
}
$(".model-box").hide()
$("#" + id).show()
this[id].execute()
},
legendlist: function () {
var mapLayerService = L.KqGIS.mapLayerService(legendlistUrl)
var params = new KqGIS.Map.GetLegendParams({
```

```
"layerIds": [2,4,5],
"format": "JSON",
"width": 40,
"height": 20
})
mapLayerService.getLegend(params).then(response => {
var result=response.result.result
var $box=$('<div id="legendlist"></div>')
var $html='<h4>图层图例</h4>'
for (var i=0,item; item=result[i++];) {
for (var j=0,childItem; childItem=item.items[j++];) {
if (childItem.data) {
$html += '<div class="item">' +
'<img src="data:image/png;base64,' + childItem.data + '">' +
'<label>' + childItem.legendName + '</label>' +
'</div>';
}}}
$box.html($html)
$('body').append($box)})
},
```

运行效果如图 4-4 所示。

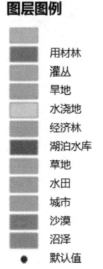

图 4-4 地图服务图例信息效果预览

4.2.2 地图服务元数据信息

以上部分完成了地图图例的获取与展示,下面继续介绍地图服务元数据信息获取流程。

第一步,初始化 kanqmapinfo。设置名称及公共执行函数。该函数主要用于对各图层及图层信息进行初始化。设置 infoModel_kanqmapinfo,包括 name、center、proj4、dpi 等多个属性,并将 infoModel_kanqmapinfo 绑定在 kanqmapinfo_pane 中。使用 changelayers 设置china 为初始展示界面。

```
kanqmapinfo: {
name: 'kanqmapinfo',
execute: (function () {
var jingjinLayer, chinaLayer;
var jingjinInfo, chinaInfo, infoModel_kanqmapinfo;
function renderInfoModel_kanqmapinfo() {
infoModel_kanqmapinfo=kendo.observable({
name: 'name',
center: '0 0',
proj4: 'proj4',
maxscale: 'maxscale',
minscale: 'minscale',
dpi: 'dpi',
unit: 'unit',
wkt: 'wkt'
});
kendo.bind($("#kanqmapinfo_pane"),infoModel_kanqmapinfo);
changeLayers_kanqmapinfo('china');
$("#loading").hide();
}
```

第二步,使用 info.getmapname 方法获取当前服务信息的名称,将其设置为 name。用同样的方法,查询当前服务信息的 proj4、wkt、center、unit 等各属性,使用 set 方法将其设置为 infoModel_kanqmapinfo 中对应的值。添加 onchange 函数,使得 infoModel_kanqmapinfo 值能够随着 selector 选择器(下拉列表)的值而变化。

```
infoModel_kanqmapinfo.set("name",info.getMapName());
infoModel_kanqmapinfo.set("proj4",info.getProj4());
infoModel_kanqmapinfo.set("wkt",info.getWkt());
var center=info.getCenter();
```

```
infoModel_kanqmapinfo.set("center",Number(center[0]).toFixed(3) + ',' + Num-
ber(center[1]).toFixed(3));
    infoModel_kanqmapinfo.set("unit",info.getUnits());
    var tileInfo=info.getTileInfo();
    var lods=tileInfo.lods;
    infoModel_kanqmapinfo.set("maxscale",lods[0].scale);
    infoModel_kanqmapinfo.set("minscale",lods[lods.length - 1].scale);
    infoModel_kanqmapinfo.set("dpi",tileInfo.dpi);}
    function onChange_kanqmapinfo() {
    var value= $ ("#kanqmapinfo_selector").val();
    changeLayers_kanqmapinfo(value);
};
```

第三步,返回时,调用函数。创建图层组,创建变量 chinainfo 承接发布的中国地图服务的 url 地址,并调用 4.2.1 中第一步的方法将其初始化。使用 singleMapLayer 添加 1～6 共 6 个图层,设置相应参数。将 kanqmapinfo_selector 绑定为 kendoDropDownList,其中 text 标题为中国行政区,值为 china。设置 show 函数,把 this.name 作为参数传入 4.2.1 第三步 _lists 中的 commonExecute 公共执行函数。同样地,设置 render 函数,使用 this.name 调用 _lists 的 render 函数。

```
return function () {
layergroup=L.layerGroup();
layergroup.addTo(map);
chinaInfo=newKqGIS.Map.GetMapInfoService(chinaUrl);
chinaInfo.init();
//添加图层
chinaLayer=L.KqGIS.singleMapLayer(chinaUrl,
{
layerI 将 ds:"1,2,3,4,5,6",style:"default",format:"image/png",transparent:true,
});
var data=[{ text:"中国行政区",value:"china" },];
$ ("#kanqmapinfo_selector").kendoDropDownList({
dataTextField:"text",dataValueField:"value",dataSource:data,index:0,
change:onChange_kanqmapinfo
});
}})();
show: function () {
_lists.commonExecute(this.name)
```

```
},
render: function () {
_lists.render(this.name)
}},
```

运行结果如图 4-5 所示。

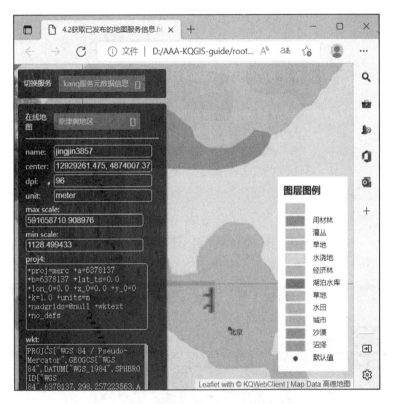

图 4-5 地图服务元数据信息效果预览

4.2.3 图层信息

图层信息的获取原理与元数据相同,参考 4.2.2 中的步骤,对图层信息进行查询。只需将 url 改为需要查询图层信息的地图服务地址,然后调用 tileMapLayer 导入图层,遍历结果并按 id 划分每个图层,用逗号隔开。

```
var mapLayerService=L.KqGIS.mapLayerService(layerinfoUrl)
mapLayerService.getLayers().then(response => {
$("#loading").css("display","none");
$("#layerinfo_result").val(formatJSON(response));
var layer=L.KqGIS.tileMapLayer(entireimageUrl,
```

```
{
layerIds：(function () {
var result=[]
for (var i=0,item; item=response[i++];) {
result.push(item.id)
}
return (result=result.join(','))
})(),
style："default",format："image/png",transparent：true,})
layer.addTo(layergroup);
})
},
show：function () {
_lists.commonExecute(this.name)
},
render：function () {
_lists.render(this.name)
}}}
```

运行效果如图 4-6 所示。

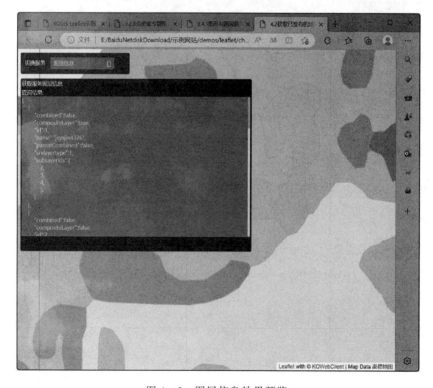

图 4-6　图层信息效果预览

4.3 获取数据库中某图层信息

为了获取数据库中的某图层信息,首先需要在本地创建数据库,然后参考 2.3 对示例数据库进行配置,在配置完成后,再进行以下步骤以完成数据库图层信息的在线获取。

4.3.1 获取字段属性

以获取字段属性为例,具体操作如下。

第一步,设置待查询的要素类名称及值,将其存入 data 数组。新建 featureclass 要素集类型变量,初始化其值为 basemap_l。设置变量 options 以储存各类参数,其内容包括:request 请求、对应地图服务实例界面中地图服务的获取字段元信息(图 4-7)、本地数据库路径、要素集名称及数据库类型。然后使用 JSON.stringify 方法将 options 的值序列化为 JSON 字符串。

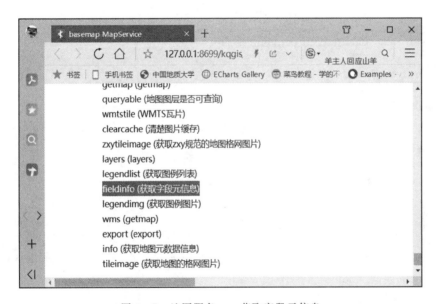

图 4-7 地图服务——获取字段元信息

```
function onload() {
var data=[
{text:"BaseMap_L",value:"BaseMap_L"},
{text:"Neighbor_P",value:"Neighbor_P"},
{text:"Landuse_R",value:"Landuse_R"},
{text:"MapDivision_L",value:"MapDivision_L"},
{text:"Captital_P",value:"Captital_P"},
];
```

```
var featureClass="BaseMap_L";
var options={
request:"GetFieldInfo",
//    dbName:"D:/KQGIS Server/jingjin/BaseMap_L" + ".shp",
dbName:dbName:"D:\\BaiduNetdiskDownload\\KQGIS示例地图V2_0514\\示例地图0514\\配图成果\\示例数据.kqdb\\行政区和界线"+".shp",
FeatureClassName:"BaseMap_L",
dbType:21,
};
$("#options").val(formatJSON(JSON.stringify(options)));
```

第二步，将 selector 绑定为"kendoDropDownList"下拉列表，设置选项文本、选项卡对应的值，当选项切换时，调用函数，读取当前选项的 index 并将其值赋给 dataitem。然后参考上一步设置 options，其中 dbname 的值由 dataitem 的 value 决定。

```
var selector=$("#selector").kendoDropDownList({
dataTextField:"text",
dataValueField:"value",
dataSource:data,
index:0,
select:function(e){
var dataItem=this.dataItem(e.item.index());
options={
request:"GetFieldInfo",
//    dbName:"D:/KQGIS Server/jingjin/" + dataItem.value + ".shp",
dbName:"D:\\BaiduNetdiskDownload\\KQGIS示例地图V2_0514\\示例地图0514\\配图成果\\示例数据.kqdb"+dataItem.value + ".shp",
FeatureClassName:dataItem.value,
dbType:21,
};
$("#options").val(formatJSON(JSON.stringify(options)));
}
}).data("kendoDropDownList");
```

第三步，初始化地图，添加高德地图矢量图层。设置用户高宽及基础高宽并计算 offset。

第四步，参考4.1，为 datasource‐directory 绑定"kendoButton"按钮，点击时执行异步函数，为 notification 绑定 kendoNotification 消息提示，设置各类参数。

第五步，使用 getParamString 解析参数并将结果及服务地址 url 作为参数输入 httpPostAsync 函数中。新建 response 变量承接函数结果。

```
//请求数据
var url='http://127.0.0.1:8699/kqgis/rest/services/basemap/map';
let response=JSON.parse(await KQCommon.NetworkTools.httpPostAsync(url,KQCommon.NetworkTools.getParamString(options)));
```

第六步,若response的结果为success,则调用回调函数,提示"获取字段属性成功",并使用JSON.stringify将response序列化为JSON数组,经formatJSON格式化解析后展示在result中。若结果为其他,则提示"获取字段属性失败"。

```
if (response.resultcode == "success") {
notification.show({message:"获取字段属性成功。"},"info");
$("#result").val(formatJSON(JSON.stringify(response)));
} else {
notification.show({message:"获取字段属性失败。"},"info");
}
}
}).data("kendoButton");
}
</script>
```

运行效果如图4-8所示。

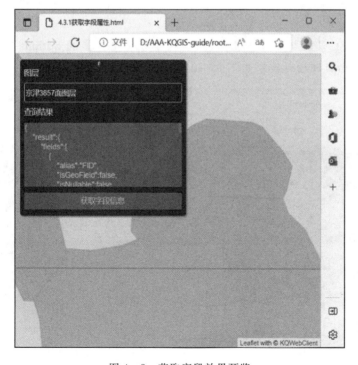

图4-8 获取字段效果预览

4.3.2 获取要素类

获取要素类的操作步骤与 4.3.1 相似,只需在 4.3.1 第一步的 options 中将 request 改为 "GetFeatureClass"即可,具体运行效果如图 4-9 所示。

图 4-9 获取要素类效果预览

4.3.3 获取要素

获取要素的操作步骤与 4.3.1 相似,只需在 4.3.1 第一步的 options 中将 request 改为 "GetFeature"即可,具体运行效果如图 4-10 所示。

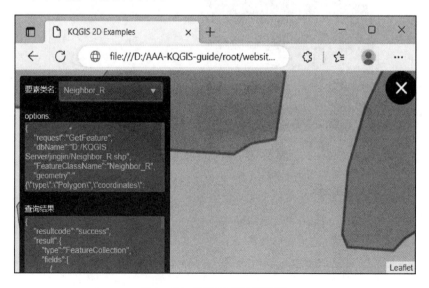

图 4-10 获取要素效果预览

5 KQGIS 数据服务

5.1 瓦片地图加载

瓦片地图金字塔模型是一种多分辨率层次模型,从瓦片金字塔的底层到顶层,分辨率越来越低,但表示的地理范围不变。KQGIS 提供了 tileMapLayer 类来实现地图文档的存放和管理。通常要显示瓦片地图需要进行如下操作:第一步,在本地使用 GIS 软件制作瓦片地图,或在地理数据平台上下载已有的瓦片地图文件,参考 2.1 将文件保存到本地数据库中;第二步,使用 KQGIS Server 将瓦片地图发布为地图服务,并在 KQGIS Services Directory 中获取其访问路径;第三步,在前端网页中调用地图服务访问接口,实现地图加载。瓦片地图的加载方式与 3.1 中提到的地图加载方式基本一致,但在图层加载之前需要使用 GetMapWMTSInfoService 类的 init 功能获取服务信息,然后正常使用 tileMapLayer 加载出图地址即可。

```
new KqGIS.Map.GetMapWMTSInfoService(url).init();
```

运行效果如图 5-1 所示。

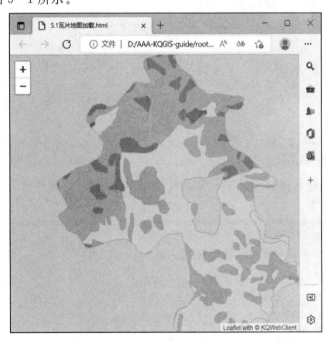

图 5-1 瓦片地图加载效果预览

5.2 矢量地图文档加载

矢量地图文档是地图文档的常见组织形式，其原理是用坐标点的方式存储点线面数据。该方式在节约存储空间的同时，还保证了要素的分辨率，使得图像在放大缩小时均不会失真。其加载策略与瓦片地图文档一致，下面重点介绍矢量地图文档中的图层动态注记及要素过滤显示。

5.2.1 图层动态注记

图层动态注记和 3.5 中的信息标注类似，但不同的是，信息标注通常由 GIS 软件生成或事先设置好，而动态注记则由网站前端用户编辑修改，并将结果实时同步更新在客户端，一般分为以下几个步骤。

第一步，初始化地图，设置地图中心点，缩放层级。设置投影为天地图 WGS84 坐标系，使用 L. layergroup 类创建图层组管理各图层，将创建好的图层组 layergroup 使用 addTo 方法加入地图。

第二步，使用 L. kqGIS. tiandituTileLayer 类初始化 vec 矢量图层及 cva 矢量注记图层，使用 addTo 方法将图层加入地图。

第三步，创建 chinaUrl 变量，地址为苍穹云发布的 wms 全国县级行政区矢量图层。把 chinaUrl 作为参数使用 L. kqGIS. wmsLayer 创建矢量图层 china4490Layer，设置展示的图层范围为 roads 图层，设置颜色及版本信息后，加入地图。

```
var chinaUrl = kqcloud. kqcloud_ip + '/kqcloudserver/wms';
//添加 wms 图层    全国县级行政区
china4490Layer=L. KqGIS. wmsLayer(chinaUrl,{
layers: kqcloud. roads,
styles: 'L -线-蓝',
version: '1.1.0'
}). addTo(map);
```

第四步，创建 changeStyles 更改样式注记函数。当传入参数 value 时，将当前 china4490Layer 设置为 value 对应的图层。调用 getLegendGraphic 函数并调用 notification 的 show 方法提示更改样式信息。参考 4.1 设置 kendoUI 容器，创建一个下拉选择框和一个信息提示框。

```
function changeStyles(value) {
china4490Layer. setParams({ styles: value });
getLegendGraphic(value);
notification. show({ message: '更改样式！' },'info');
}
$ ('#legend'). kendoListView({
```

```
dataSource: [],
template: kendo.template($('#javascriptTemplate').html())
});
getLegendGraphic('L-线-蓝');
var china4490Layer,map;
var notification=$('#notification')
.kendoNotification({
position: {
pinned: true,
bottom: 12,
right: 12
},
autoHideAfter: 2000,
stacking: 'up',
templates: [
{
type: 'info',
template: <div># = message #</div>
}
]
})
.data('kendoNotification');
}
```

第五步,创建 getLegendGraphic 函数,当传入参数 value 时,调用 KqGIS.Bigdata.Query.LegendParams 类返回查询参数,并使用 params 变量承接。随后设置 params 参数的 style,这一步可以修改为设定 params 的其他属性,从而达到对动态注记的各种渲染。

KqGIS.Bigdata.Query.LegendParams 类的参数如表 5-1 所示。

表 5-1 KqGIS.Bigdata.Query.LegendParams 参数介绍

Name	Type	Default	参数定义
url	string		服务全地址
version	string	1.1.0	可选,服务版本号,平台支持的版本为1.1.0
request	string	GetLegendGraphic	可选,请求的操作名称,这里的值为 GetLegendGraphic
strict	boolean	false	可选,必须参数
style	string		样式名称
format	string	application/JSON	可选,输出格式:application/JSON

续表 5-1

Name	Type	Default	参数定义
legendOptions	string	forceLabels:on	可选,控制是否返回 label:forceLabels:on(on 为返回,false 为不返回,默认返回)。控制 label 的字体、颜色、大小:fontColor:0x0012faff(默认为黑色);fontSize:12(默认为 12);bgColor:0x0012faff(默认为白色)。多个条件时,用英文分号隔开。颜色只能为 16 位的颜色码
customResponse	boolean	true	可选,必须参数
transparent	boolean	false	可选,背景透明,默认为 false
width	number	20	可选,图片的宽度,默认为 20(以像素为单位)
height	number	20	可选,图片的高度,默认为 20(以像素为单位)

```
function getLegendGraphic(value) {
var params=new KqGIS.Bigdata.Query.LegendParams();
params.style=value;
```

随后将 kqcloud_ip 云服务地址传入 L.KqGIS.Bigdata.queryService 类中创建 query 对象,调用对象的 legend 方法。如果查询成功,则令常量 legends 等于查询到的 data 的 legend 值,并将创建的 kendo 容器中的 DataSource 设置为 legends,同时更改下拉列表中的 legend 值(表 5-2)。

表 5-2 legend 方法参数介绍

Name	Type	参数定义
params	KqGIS.Bigdata.Query.LegendParams	查询需要的相关参数类
onSuccess	RequestCallback	成功回调函数
onFailed	RequestCallback	失败回调函数

```
var query=L.KqGIS.Bigdata.queryService(kqcloud.kqcloud_ip);
query.legend(params,function (data) {
if (data) {
const legends=data.result.Legend;
const dataSource=new kendo.data.DataSource({
data: legends
});
const listView= $ ('#legend').data('kendoListView');
listView.setDataSource(dataSource);
}
});
}
```

第六步,设置 onchange 函数,读取并将 selector 中的值赋给 value,随后调用 changestyle 函数。设置 data 数组,包括蓝线、红线、公路线 3 种风格,供用户切换。最后设置 selector,绑定 kendoui 的下拉列表,数据源为 data,文本域设置为 text,值域设置为 value,添加 onchange 方法。

```
function onChange() {
var value = $('#selector').val();
changeStyles(value);
}
var data = [
{ text: '蓝', value: 'L-线-蓝' },
{ text: '红', value: 'L-线-红' },
{ text: '公路线-简约风', value: 'L-线-公路线-简约风' }
];
$('#selector').kendoDropDownList({
dataTextField: 'text',
dataValueField: 'value',
dataSource: data,
index: 0,
change: onChange
});
```

运行效果如图 5-2 所示。

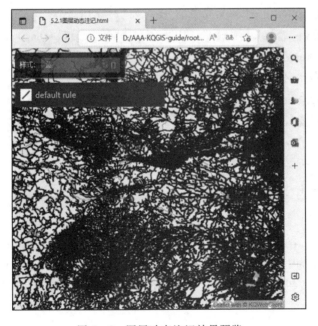

图 5-2　图层动态注记效果预览

5.2.2 要素过滤显示

要素过滤显示用于在矢量图层中分别展示不同要素的情况或根据条件筛选要素。其实现方法分为以下几步。

第一步，设置查询条件数组，可以预先设置或动态设置，这里以预先设置为例。参考4.1.1 添加 kendo 容器的下拉列表，设置数据源、文本等信息。

```
function onload() {
var data=[
{ text："fclass"='primary_link',value："fclass"='primary_link' },
{ text："code"=5142',value："code"=5142' },
{ text："oneway" ='F',value："oneway" ='F' }
];
$("#property").kendoDropDownList({
dataTextField："text",
dataValueField："value",
dataSource：data,
index：0
});
```

第二步，初始化地图，添加图层组。将天地图 vec 和 cva 图层加入图层组。使用 L.rectangle 创建矩形要素并加入地图。参考5.2.1添加苍穹云的 WMS 服务并添加 roads 图层。

```
 L.rectangle([[27.825896790055133,107.14272816540743],[33.07736163380513,
115.29458363415743]],{ color："red",fillOpacity：0 }).addTo(map);
var chinaUrl=kqcloud.kqcloud_ip + "/kqcloudserver/wms";
var wmsLayer=L.KqGIS.wmsLayer(chinaUrl,{
layers：kqcloud.roads,
styles："L-线-蓝",
useindex：false,// 当前样式若已经缓存，则需要加上该参数，否则看不到过滤效果
version："1.1.0"
}).addTo(map);
```

第三步，参考4.1.1 创建 kendo 容器，首先创建"query"按钮用于执行查询功能，创建 spacefilter 承接 space 容器的值与当前几何要素相交查询的结果，propfilter 承接 property 属性的值，cql_filter 作为二者的合集，使用 wmsLayer.setParams 设置查询参数为 cql_filter。然后创建 reset 复位按钮，点击后将 cql_Filter 的值设置为"1=1"初始值。最后创建信息窗口，设置其样式风格。

```
$("#query").kendoButton({
click: function () {
var spaceFilter='( INTERSECTS(the_geom,' + $("#space").val() + ') )';
var propfilter= $("#property").val();
var cql_filter=spaceFilter + ' and (' + propfilyer + ')'
wmsLayer.setParams({ cql_Filter: cql_filter });
}
}).data("kendoButton");;
$("#reset").kendoButton({
click: function () {
wmsLayer.setParams({ cql_Filter: '1=1' });
}
}).data("kendoButton");
var notification= $("#notification").kendoNotification({
position: {
pinned: true,
bottom: 12,
right: 12
},
autoHideAfter: 2000,
stacking: "up",
templates: [{
type: "info",
template: <div># = message #</div>
}]
}).data("kendoNotification");
};
```

运行效果如图5-3所示。

图 5-3 要素过滤显示效果预览

5.3 矢量瓦片加载

矢量瓦片类似栅格瓦片,是将矢量数据用多层次模型分割成矢量要素描述文件存储在服务器端,然后传输到客户端根据指定样式渲染绘制地图,在单个矢量瓦片上存储着投影于一个矩形区域内的几何信息和属性信息。当客户端通过分布式网络获取矢量瓦片、地图标注字体、图标、样式文件等数据后,最终在客户端渲染输出地图。其实现方法如下。

第一步,初始化地图、高德地图矢量图层。

第二步,获取矢量图层服务地址及出图地址,将其合并为 styleJSONurl,然后作为参数输入 options 中。使用 vectorTileLayer 接收矢量瓦片对象,以 options 作为 style 并添加至图层,其中矢量瓦片对象参数如表 5-3 所示。

表 5-3 vectorTileLayer 参数介绍

Name	Type	Default	参数定义
style	string	'style.JSON'	可选,mapbox 标准的 style
minZoom	number	0	可选,最小比例尺
maxZoom	number	18	可选,最大比例尺

```
let styleJSONUrl=service_ip+'/jingjin3857_slwp-proxy/kqgis/tilecache/jingjin3857_
slwp/style.JSON? ua_token='+server_token;
let options={
style：styleJSONUrl
};
L.KqGIS.vectorTileLayer(options).addTo(map);
}
```

运行效果如图 5-4 所示。

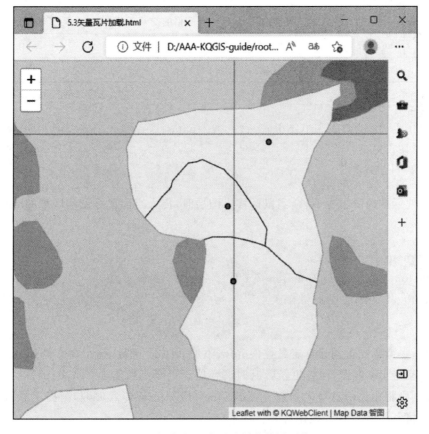

图 5-4　矢量瓦片图层加载效果预览

5.4　属性瓦片加载

属性瓦片将地图中矢量图层的属性数据以属性瓦片的形式进行存储,支持 UTFGrid 格式。在地图服务的应用中,如果包含较多的鼠标交互操作,传统做法是在地图上叠加要素图层,每个要素具有自己的热点和事件,用于完成鼠标交互。但在大数据量、高并发请求的环中,客户端尤其是移动终端上,就不能很好地渲染大数据量的地理要素,因此面临严重的性能问题。这种情况下,出现了一种地图瓦片结合要素属性信息的缓存方式,也就是在传统地图瓦片的基础上,额外存储了按照格网划分的要素属性信息,这种预先划分的要素属性信息,称为属性瓦片(也称互动格网瓦片)。其具体实现步骤如下。

第一步,初始化地图、天地图 vec 和 cva 图层。

第二步,添加属性瓦片图层,其参数包括 url 服务地址以及 options,其值如表 5-4 所示。

表 5-4 utfGridLayer 参数介绍

Name	Type	Default	参数定义
useJSONP	string	true	可选,启用 JSONP 方式,启用后需在 url 路径后在 callback={cb}
pointerCursor	string	true	可选,当鼠标悬停在网格的交互部分上时,将鼠标光标更改为指针
minZoom	number	0	可选,最小比例尺
maxZoom	number	18	可选,最大比例尺

```
//属性瓦片图层
let utfGrid=L. KqGIS. utfGridLayer('../../data/ecoregions/{z}/{x}/{y}.JSON',{
useJSONP: false
}).addTo(map);
```

随后创建函数监听鼠标事件,当鼠标移入图层时,传入当前鼠标悬停区域的 data 值,更新 info 信息。

```
//监听 mouseover 事件
utfGrid.on('mouseover',function (e) {
info.update(e.data);
});
```

第三步,参考 3.4.1 创建自定义控件 info,添加 onAdd 函数,运行时会调用 domutil 方法自动为 info 创建 div 容器,检查更新后返回该容器。更新函数需要传入 data 作为参数,若成功传入 data,则将相关信息写入当前容器的 innerHTML;若传入失败,则提示悬停在行政区查看信息。将 info 控件加入地图。

```
//创建信息展示控件
let info=L. control();
info.onAdd=function (map) {
this._div=L. DomUtil.create('div','info');
this.update();
return this._div;
};
```

```
info.update=function (data) {
if (data) {
this._div.innerHTML =
'<h4>中国行政区信息:</h4>' +
(data? 'NAME:' +data.NAME +'<br />' +'CODE:' +data.CODE +'<br />' +
'NAME_ENG:'+data.NAME_ENG +'<br />' +'NAME_PY:' +data.NAME_PY: '无效
数据');
} else {
this._div.innerHTML='<h4>中国行政区信息:</h4>悬停在行政区查看信息';
}
};
info.addTo(map);
```

第四步,创建 getColor 函数,传入参数 size,根据 size 大小分配颜色。highlightFeature 函数用于对选中区进行高亮显示,且让其适配各种浏览器。Style 函数传入 featrue,设置其初始样式,并调用 getColor 函数设定颜色。

```
function getColor(size) {
return size > 5000? '#EE1289': size > 2500? '#EE3B3B': size > 2000? '#EE3B3B':
size > 1500? '#EEA2AD'
: size > 1000? '#EECBAD': size > 500? '#EED5D2': size > 100? '#EEE8CD': '#
F0F0F0';
}
function style(feature) {
return {
weight: 2,
opacity: 1,
color: 'white',
dashArray: '3',
fillOpacity: 0.7,
fillColor: getColor(feature.properties.size)
};
}
function highlightFeature(e) {
var layer=e.target;
layer.setStyle({
weight: 5,
```

```
color: '#666',
dashArray: '',
fillOpacity: 0.7
});
if (! L.Browser.ie && ! L.Browser.opera && ! L.Browser.edge) {
layer.bringToFront();
}}
```

第五步，接上一步，先新建 geoJSON 创建 resetHighlight 函数用于对 geoJSON 目标进行样式设置。创建 zoomToFeature 函数，当鼠标事件触发时调用 map.fitbounds 函数将屏幕显示范围设置为所选区域。创建 onEachFeature 函数，传入 feature 和 layer 两个参数，当鼠标移入图层时，执行 highlightFeature；移出图层时，执行 resetHighlight；点击图层时，执行 zoomToFeature。

```
var geoJSON;
geoJSON=L.geoJSON(china,{
style: style,
onEachFeature: onEachFeature
}).addTo(map);
function resetHighlight(e) {
geoJSON.resetStyle(e.target);}
function zoomToFeature(e) {
map.fitBounds(e.target.getBounds());
}
function onEachFeature(feature,layer) {
layer.on({
mouseover: highlightFeature,
mouseout: resetHighlight,
click: zoomToFeature
});
}}
</script>
```

运行效果如图 5-5 所示。

图 5-5 属性瓦片效果预览

5.5 第三方地图加载

第三方地图主要是指互联网上涌现的大量地图服务,提供免费开放的基础地图服务,一般均为瓦片地图形式,常在应用中作为底图使用。网络上主流的地图服务有天地图、高德地图、智图、OSM 地图、百度地图、腾讯地图、谷歌地图、必应地图、ArcGIS 在线地图和 Mapbox 地图等。

这些免费的在线地图服务吸引了众多用户,方便了广大开发者使用在线地图开发丰富的地图应用,挖掘 GIS 的潜在价值。KQGIS Client for Leaflet 对这些互联网地图信息进行了封装,提供一套通用的地图加载方法。

5.5.1 OSM 地图加载

具体实现代码如下。

第一步,创建待添加图层名变量,创建 changLayer 函数,传入 value 值后,调用 clearLayers 清除当前图层,然后根据 value 值使用 addTo 方法将对应图层加入地图。

```
<script type='text/javascript'>
let normalLayer,bikeLayer,transportLayer,humanitarianLayer,map,layerGroup;
function changeLayers(value) {
layerGroup.clearLayers();
if (value === 'normal') {
```

```
normalLayer.addTo(layerGroup);
} else if (value === 'bike') {
bikeLayer.addTo(layerGroup);
} else if (value === 'transport') {
transportLayer.addTo(layerGroup);
} else if (value === 'humanitarian') {
humanitarianLayer.addTo(layerGroup);
}
}
```

第二步,创建 onChange 函数,当 selector 的值变化时,调用 changeLayer 方法。创建 on-load 函数,初始化 data 数组,设定待切换地图的名称和值,参考 4.1.1 创建"kendo"下拉列表,绑定 selector 容器,设置数据源为 data、方法为 onChange,设置文本域及值域的值。

```
function onChange() {
let value= $('#selector').val();
changeLayers(value);
}
function onload() {
let data=[
{ text: 'OSM 标准',value: 'normal' },
{ text: 'OSM 自行车地图',value: 'bike' },
{ text: 'OSM 交通地图',value: 'transport' },
{ text: 'OSM 人道主义地图',value: 'humanitarian' }
];
$('#selector').kendoDropDownList({
dataTextField: 'text',
dataValueField: 'value',
dataSource: data,
index: 0,
change: onChange
});
```

第三步,初始化地图 map,设置地图中心、缩放等级。创建图层组,使用 addTo 方法加入地图。使用 L.KqGIS.osmTileLayer 创建多个图层对象。调用 changeLayers 函数设置默认 normal 图层为初始图层。

```
layerGroup=L.layerGroup();
layerGroup.addTo(map);
```

```
normalLayer=L. KqGIS. osmTileLayer({ layerType：'normal' });
bikeLayer=L. KqGIS. osmTileLayer({ layerType：'bike' });
transportLayer=L. KqGIS. osmTileLayer({ layerType：'transport' });
changeLayers('normal');
```

运行效果如图 5-6 所示。

图 5-6　OSM 地图加载效果预览

5.5.2　天地图加载

天地图的加载方式与 5.5.1 中的相同,使用时只需更改 L. KqGIS. tiandituTileLayer 来实例化对象,确定 layerType 即可。

```
layergroup=L. layerGroup();
layergroup. addTo(map);
vecLayer=L. KqGIS. tiandituTileLayer({ layerType：'vec' });
cvaLayer=L. KqGIS. tiandituTileLayer({ layerType：'cva' });
evaLayer =L. KqGIS. tiandituTileLayer({ layerType：'eva' });
```

运行效果如图 5-7 所示。

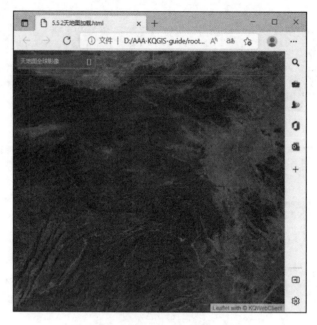

图 5-7　天地图加载效果预览

5.5.3　百度地图加载

百度地图的加载方式与 5.5.1 中的相同，使用时只需更改 L. KqGIS. baiduTileLayer 来实例化对象，确定 layerType 即可。

```
layergroup=L. layerGroup();
layergroup. addTo(map);
vecLayer=L. KqGIS. baiduTileLayer({ layerType:'vec' });
imgdLayer=L. KqGIS. baiduTileLayer({ layerType: 'imgd' });
imgzLayer=L. KqGIS. baiduTileLayer({ layerType: 'imgz' });
```

运行效果如图 5-8 所示。

5.5.4　高德地图加载

高德地图的加载方式与 5.5.1 中的相同，使用时只需更改 L. KqGIS. gaodeTileLayer 来实例化对象，确定 layerType 即可。

```
layerGroup =L. layerGroup();
layerGroup. addTo(map);
vecLayer=L. KqGIS. gaodeTileLayer({ layerGroup: 'vec' });
imgLayer=L. KqGIS. gaodeTileLayer({ layerType: 'img' });
ciaLayer=L. KqGIS. gaodeTileLayer({ layerType: 'cia' });
```

图 5-8 百度地图加载效果预览

运行效果如图 5-9 所示。

图 5-9 高德地图加载效果预览

5.6 OGC 地图服务加载

OGC 为开放地理空间信息联盟(Open Geospatial Consortium),它的主要作用是制定与空间信息、基于位置服务相关的标准。而这些标准其实就是一些接口或编码的技术文档,不同的厂商、各种 GIS 产品都可以对照这些文档来定义开放服务的接口、空间数据存储的编码、空间操作的方法。OGC 目前提供的标准多达几十种,包括我们常用到的 WMS、WFS、WCS、WMTS 等,还有一些地理数据信息的描述文档,比如 KML、SFS(简单对象描述)、GML、SLD(地理数据符号化)等。

地图服务是一种使地图可以通过 Web 访问的方法。首先需要制作原始地图,然后发布到服务站点上,用户便可以通过各种 Web 应用来访问地图相应的地图服务。

5.6.1 WMS

2.6.2 中已经详细描述了 WMS 服务的发布流程,现将 WMS 服务前端加载流程介绍如下。

第一步,初始化各 WMS 图层,创建 changeLayers 函数实现图层切换,传入参数 value,清空图层组当前图层后,根据 value 的值将对应图层加入图层组,并使用 setView 方法调整地图中心及缩放等级。创建 onChange 函数,当 selector 触发时,将其值赋给 value,并调用 changeLayer 函数。

```javascript
<script type='text/javascript'>
let osmWmsLayer, osmOverlayWmsLayer, topoOsmWmsLayer, china3857Layer,
jingjin3857Layer,map,layergroup;
function changeLayers(value) {
layergroup.clearLayers();
if (value === 'osm-wms') {
osmWmsLayer.addTo(layergroup);
} else if (value === 'osm-overlay-wms') {
osmOverlayWmsLayer.addTo(layergroup);
} else if (value === 'topo-osm-wms') {
topoOsmWmsLayer.addTo(layergroup);
} else if (value === 'china3857') {
china3857Layer.addTo(layergroup);
map.setView([30.543,114.397],4);
} else if (value === 'jingjin3857') {
jingjin3857Layer.addTo(layergroup);
```

```
map.setView([40,116.3],7);
    }
}
function onChange() {
    let value=$('#selector').val();
    changeLayers(value);
}
```

第二步,创建 onload 函数,初始化各图层信息,参考 4.1.1 创建下拉列表,设定数据源等信息。

```
function onload() {
    let data=[
        {text: '中国行政区',value: 'china3857'},
        {text: '京津冀地区',value: 'jingjin3857'},
        {text: 'TOPO-OSM-WMS',value: 'topo-osm-wms'},
        {text: 'OSM-WMS',value: 'osm-wms'},
        {text: 'OSM-Overlay-WMS',value: 'osm-overlay-wms'},
    ];
    $('#selector').kendoDropDownList({
        dataTextField: 'text', dataValueField: 'value', dataSource: data, index: 0, change: onChange
    });
}
```

第三步,初始化地图 map 及地图信息,缩放层级。使用 addTo 方法,将图层组加入 map 中,初始化 WMS 出图地址。

```
//添加 wms tile 图层
let url ='http://ows.mundialis.de/services/service?';let china3857Url=service_ip+(isMicroService?'/china3857-proxy':'')+'/kqgis/rest/services/china3857/map/wms';let jingjin3857Url=service_ip+(isMicroService?'/jingjin3857-proxy':'')+'/kqgis/rest/services/jingjin3857/map/wms';
```

使用 WMSLayer 创建各图层,WMSLayer 参数除了 url 服务地址之外,还有 options 列表,其值如表 5-5 所示。

表 5-5 WMSLayer 参数介绍

Name	Type	Default	参数定义
layers	string		图层名称
styles	Object		可选,图层样式
format	string	'image/png'	可选,图像格式
transparent	string	true	可选,服务返回的图像是否透明
version	string	'1.1.1'	可选,版本信息
crs	L. Proj. CRS	null	可选,坐标参考系统。如果你不确定它的意思,不要更改它

运行效果如图 5-10 所示。

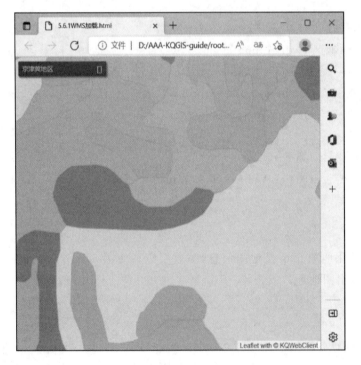

图 5-10 WMS 图层加载效果预览

5.6.2 WMTS

2.6.1 中已经详细描述了 WMTS 服务的发布流程,WMTS 图层的加载方法与 5.6.1 中的基本一致,而 WMTSLayer 参数除了 url 服务地址之外,还有 options 列表,其值如表 5-6 所示。

5 KQGIS 数据服务

表 5-6　WMTSLayer 参数介绍

Name	Type	Default	参数定义
layer	string		可选,图层名称
style	Object		可选,图层样式
maxZoom	Number		可选,图层最大的比例尺
scales	Array		可选,服务的比例尺数组,默认为和地图相同
resolutions	Array		可选,服务的分辨率数组,默认为和地图相同。优先使用分辨率,更精确
fixParamNames	Map.<Array.<Array.<String>>>		可选修正服务名。new Map([['tilematrix','z'],['tilerow','y'],['tilecol','x'],['style','styleId']])
crs	L. Proj. CRS		可选,服务的投影对象,默认为和地图相同
origin	Array.<number>|L. Point		可选,服务的投影原点,默认为和地图相同
opacity	Number		可选,透明度
format	string	'image/png'	可选,wmts 图像格式
tileSize	number|L. Point	'256'	可选,瓦片大小
tilematrixSet	Object		可选,瓦片矩阵集
version	string	'1.0.0'	可选,版本信息
attribution	string		可选,版权信息

使用 WMTSLayer 存储图层,具体形式如下。

```
worldStreetMapLayer=L.KqGIS.wmtsLayer(worldStreetMapUrl,
{
layer:[0],
style:'default',
tilematrixSet:'default028mm',
format:'image/png'
}
);
worldImageryLayer =L.KqGIS.wmtsLayer(worldImageryUrl,
{
layer:[0],
style:'default',
tilematrixSet:'default028mm',
format:'image/png'
}
);
```

```
china3857Layer=L.KqGIS.wmtsLayer(china3857Url,
{
layer: [1,2,3,4,5,6],
style: 'default',
tilematrixSet: 'default028mm',
format: 'image/png'
}
);
jingjin3857Layer=L.KqGIS.wmtsLayer(jingjin3857Url,
{
layer: [1,2,3,4,5,6,7],
style: 'default',
tilematrixSet: 'default028mm',
format: 'image/png'
}
);
changeLayers('china3857');
}
</script>
```

运行效果如图 5-11 所示。

图 5-11　WMTS 图层加载效果预览

5.6.3 WFS

2.6.3中已经详细描述了WFS服务的发布流程,现将WFS服务前端加载流程介绍如下。

第一步,初始化地图,设置地图中心和缩放层级。初始化图层组,添加高德地图作为底图,设置WFS服务地址service_url。

```
let service_url = service_ip + (isMicroService ? '/jingjin_data-proxy': ") + "/kqgis/rest/services/jingjin_data/data/wfs";
```

第二步,以service_url为参数,使用L.KqGIS.dataWFSService实例化对象wfsinfo。使用KqGIS.Data.WFSInfoParams实例化对象getCapabilitiesParams。其中Request代表请求的类型,可能的值为:"GetCapabilities"——返回WFS服务支持的能力描述信息;"DescribeFeatureType"——返回要素的结构信息;"GetFeature"——框查询要素信息。这3种情况下request参数必选。

```
let wfsinfo = L.KqGIS.dataWFSService(service_url);
let getCapabilitiesParams = new KqGIS.Data.WFSInfoParams(
{
request: "GetCapabilities"
}
);
let typenames = null;
```

第三步,使用dataWFSInfo方法返回WFS服务支持的能力描述信息,用xml格式表示,解析获取相关的参数。首先,使用DOMParser初始化对象domParser,将DOM树转换为XML或HTML源。然后,调用parseFromString方法解析dom并转为string,使用xmlDoc承接,把xmlDoc对象的子节点转化为内文本并赋值给typenames。最后调用WFSInfoParams进行查询,该参数除了url服务地址之外,还有options列表,其值如表5-7所示。

表5-7 WFSInfoParams参数介绍

Name	Type	参数定义
request	string	可选,请求的类型表示。可能的值为:"GetCapabilities"——返回WFS服务支持的能力描述信息;"DescribeFeatureType"——返回要素的结构信息;"GetFeature"——框查询要素信息。这3种情况下request参数必选
service	string	可选,服务类型,固定为"WFS"
typename	string	可选,要素类型列表与参数typenames互斥,WFS 1.0.0、WFS 1.1.0专有参数,必填

续表 5-7

Name	Type	参数定义
typenames	string	可选,要素类型列表与参数 typename 互斥,WFS 2.0.0 专有参数,必填
filter	string	可选,框选矩形。通过矩形框查询要素信息时填写
bbox	string	可选,框选矩形。通过矩形框查询要素信息时填写
version	string	可选,WFS 版本号,默认为 2.0.0,支持:1.0.0,1.1.0,2.0.0。不同版本号,body 参数的结构可能有所不同,请参考 WFS 各版本的说明
body	Object	可选,body 数据。GetCapabilities 时此参数为空;DescribeFeatureType 或 GetFeature 时,数据格式为"application/json";要素插入、要素更新、要素替换、要素删除时,数据格式为"application/xml"

其中 request 为"GetFeature",typename 则为刚刚创建的内文本信息。

```
wfsinfo.dataWFSInfo(getCapabilitiesParams).then((response) => {
//返回 WFS 服务支持的能力描述信息,用 xml 格式表示,解析获取相关的参数
let domParser=new DOMParser();
let xmlDoc=domParser.parseFromString(response,'text/xml');
typenames=xmlDoc.children[0].children[3].children[2].children[1].innerHTML
}).then(() => {
let getFeatureParams=new KqGIS.Data.WFSInfoParams(
{
request:"GetFeature",
typenames:typenames
}
);
```

第四步,同第三步一样,使用 getFeatureParams 解析返回的 xml 获取要素进行展示。如果子节点 children[0]长度大于零,则对子节点下所有要素进行遍历并使用 marker 类添加标记和悬浮名称。

```
wfsinfo.dataWFSInfo(getFeatureParams).then((result) => {
let domParser=new DOMParser();
let xmlDoc=domParser.parseFromString(result,'text/xml');
if (xmlDoc.children[0].children.length > 0) {
for (var datalength=0; datalength < xmlDoc.children[0].children.length; datalength++) {
var data=xmlDoc.children[0].children[datalength].children[0].children[1].children[0].children[0].textContent.split(" ");
```

```
//添加标记
L.marker([Number(data[0]),Number(data[1])],{
//添加悬浮名称
title:xmlDoc.children[0].children[datalength].children[0].children[6].innerHTML
}).addTo(map);
```

运行效果如图 5-12 所示。

图 5-12　WFS 服务加载效果预览

6 KQGIS 要素服务

6.1 查询要素

要素查询大体上可分为几何查询和属性查询两部分,几何查询是对查询范围内的几何要素进行查询,包括点、线、面(矩形、简单多边形、带洞多边形等),通常由用户手动输入或用鼠标划定查询范围,系统查询后返回各类要素的具体信息(坐标点、样式、属性等)。

属性查询则是通过设定某些筛选条件,根据条件和所选范围来确定满足条件的要素,并返回要素的具体信息。

如需实现要素查询功能,需要在地图文档中创建要素图层并上传至 KQGIS Server 中,具体操作如下:

第一步,打开 KQGIS Desktop,点击"文件"→"打开工程"(图 6-1)。

图 6-1 KQGIS Desktop 主界面

第二步,打开左侧的内容视窗,点击左上角加号→"新建组"→"加载数据"→"加载矢量图层",点击需要发布的图层进行添加(图 6-2)。

图 6-2 KQGIS Desktop 加载数据界面

第三步,参考 2.4 进行服务发布,发布完成之后进入 http://127.0.0.1:8699/kqgis/rest/services 对服务进行查看,在 Layers 属性中可以看到已发布的图层及对应的 ID(图 6-3)。

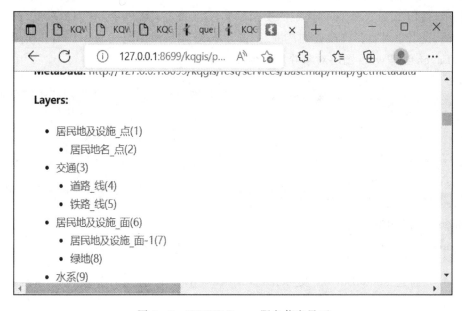

图 6-3 KQGIS Server 服务信息界面

6.1.1 点、线、面查询

参考 6.1 上传要素图层之后,可以在 Server 中浏览不同要素对应的图层号。现将查询指定范围内的要素具体信息的核心代码展示如下:

第一步,创建 data 数组,用于存放查询条件信息,创建 kendoDropDownList 并绑定 property 容器。将创建的 data 作为数据源,同时设置文本域和值域。初始化地图,设置地图中心、缩放等级,设定参考坐标系为天地图 WGS-84 坐标。创建图层组,使用 addTo 方法将图层组加入地图,并用同样的方法将天地图 vec、cva 图层加入图层组。

```
<script type="text/javascript">
function onload() {
var data=[
{ text:'"fclass"='primary_link'',value:'"fclass"='primary_link' },
{ text:'"code"=5142',value:'"code"=5142' },
{ text:'"oneway"='F'',value:'"oneway"='F' }
];
$("#property").kendoDropDownList({
dataTextField: "text",
dataValueField: "value",
dataSource: data,
index: 2
});
```

第二步,使用 L.rectangle 创建矩形要素对象,设置高亮,用于指示查询的范围(此处为固定的矩形框,读者也可自行修改查询范围)。设置 wms 图层出图地址 wmsUrl,并使用 L.KqGIS.wmsLayer 创建 wms 图层对象并加入地图。

```
 L.rectangle([[27.825896790055133,107.14272816540743],[33.07736163380513,
115.29458363415743]],{ color:"red",fillOpacity:0 }).addTo(map);
var wmsUrl=kqcloud.kqcloud_ip + "/kqcloudserver/wms";
var wmsLayer=L.KqGIS.wmsLayer(wmsUrl,{
layers: kqcloud.roads,
styles: "L-线-公路线-简约风",
version: "1.1.0"
}).addTo(map);
```

参考 4.1 创建 kendoButton 及 kendoNotification,将它们与之前创建的容器绑定,设置样式。

```
$("#query").kendoButton({
click: function () {
$("#result").val('');
$("#loading").css("display","block");
queryDetail()
}
}).data("kendoButton");;
$("#reset").kendoButton({
click: function() {
$("#result").val('');
}
}).data("kendoButton");;
var notification= $("#notification").kendoNotification({
position: {
pinned: true,
bottom: 12,
right: 12
},
autoHideAfter: 2000,
stacking: "up",
templates: [{
type: "info",
template: <div>#= message #</div>
}]
}).data("kendoNotification");
```

第三步,创建queryDetail函数,首先设置查询地址url,然后使用KqGIS.Bigdata.Query.QueryDetailParams创建查询对象。该类参数如表6-1所示。其中geomList和where的值分别取space和property容器中的值。Layerid的值为kqcloud苍穹云提供的roads图层的第一级子图层。

表6-1　QueryDetailParams类参数介绍

Name	Type	Default	参数定义
url	string		服务全地址
layerId	string		图层ID
geomList	Array.<string>		可选,图形数组
fields	Object		可选,返回字段

续表 6-1

Name	Type	Default	参数定义
where	string		可选,过滤条件
pageIndex	number	1	可选,分页页码,Default value：1
pageSize	number	10	可选,每页条数,Default value：10
fileType	string		可选,下载文件类型(shp,excel,csv),不为空时下载
returnGeom	boolean	true	可选,返回 geom

```
function queryDetail() {
var url=kqcloud.kqcloud_ip
var params=new KqGIS.Bigdata.Query.QueryDetailParams();
params.geomList=[$("#space").val()]
params.layerId =   kqcloud.roads.split(':')[1];
params.where = $("#property").val();
```

第四步,以 url 为参数,调用 queryService 方法。该方法首先启动"loading"加载框,然后调用 querydetail 函数得到返回值 result,如果该值不为 200,则返回。并将信息展示在创建的展示框中,在信息框中提示"查询成功",否则提示"查询失败"。

```
L.KqGIS.Bigdata.queryService(url)
.queryDetail(params,(response)=>{
$("#loading").css("display","none");
response=response.result;
if(response.code!==200){
return;
}
$("#result").val(formatJSON(response));
notification.show({
message："查询成功!"
},"info");
},(err)=>{
$("#loading").css("display","none");
notification.show({ message："查询失败!" },"info");
```

运行效果如图 6-4 所示。

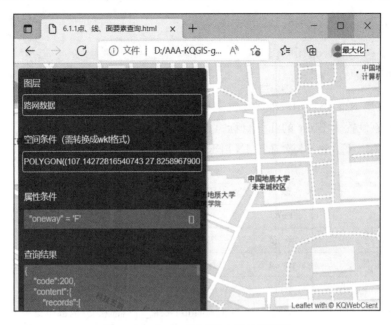

图 6-4 点、线、面要素查询效果预览

6.1.2 多边形查询

多边形是 GIS 要素中非常重要的一类数据,通常可以将多边形粗略地分为简单多边形及带洞多边形两种。下面将介绍两种对应多边形的查询并返回要素 JSON 数据的方法。

6.1.2.1 简单多边形

简单多边形的查询,与 6.1.1 中介绍的查询过程基本一致,其中多边形初始化过程如下,首先调用 leaflet 提供的 L.geoJSON 类构造要素对象,对象类型为 polygon 多边形。设置多边形顶点坐标,设置颜色透明度等属性后,使用 addTo 方法加入地图。

```
var polygon=L.geoJSON(
{
type: 'Polygon',
coordinates: [
[
[115.36508793228916,38.312993059223494],
[118.30942386978916,38.312993059223494],
[117.30942386978916,40.257328996723494],
[115.36508793228916,38.312993059223494]
]
]
```

```
},
{ style: { color: 'red', fillOpacity: 0.1 } }
).addTo(map);
```

同样地,使用 KqGIS.FilterParam 类来构造查询条件对象 Filter,使用 L.Util 扩展方法,将 polygon 对象设置为 filter 的几何目标。设置坐标系及参照空间关系类型。其中参照空间关系类型共分为以下几种,本次选择 INTERSECT 相交作为查询条件(表 6-2)。

表 6-2　SpatialRel 空间关系类型介绍

Name	Type	Default	参数定义
DISJOINT	string	Disjoint	相离
INTERSECT	string	Intersect	相交
ENVELOPEINTERSECT	string	EnvelopeIntersect	外接矩形相交
INTERIORINTERSECT	string	InteriorIntersect	内部相交(内部相交=相交-相离)
TOUCHES	string	Touches	相接
OVERLAPS	string	Overlaps	覆盖
CROSSES	string	Crosses	穿越
CONTAINS	string	Contains	包含
WITHIN	string	Within	被包含
EQUALS	string	Equals	相等

```
let filter=new KqGIS.FilterParam({
geometry: L.Util.toKqGISGeometry(polygon),
geoSRS: geoSRS,
SpatialRel: KqGIS.SpatialRel.INTERSECT
});
```

使用要素空间查询参数类 KqGIS.Data.GetFeaturesByGeometryParams 创建查询参数对象 params,设置其图层号 layerId 为 2,查询过滤条件为 filter。是否返回要素信息为 true,是否返回要素图形信息为 true。最后使用 dataFeatureService,传入服务地址,查询参数 params,执行要素查询。

```
let params=new KqGIS.Data.GetFeaturesByGeometryParams({
layerId: 2,
filter: filter,
returnContent: true,
```

```
hasGeometry: true
});
L. KqGIS. dataFeatureService(dataUrl). getFeaturesByGeometry(params, onsuccess,
onfailed);
}
})
.data('kendoButton');
```

运行效果如图 6-5 所示。

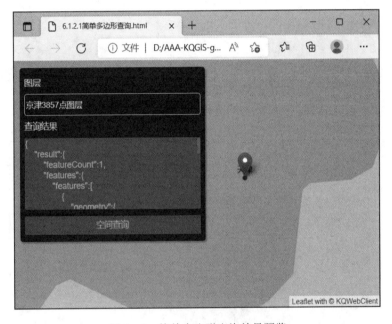

图 6-5 简单多边形查询效果预览

6.1.2.2 带洞多边形

带洞多边形,顾名思义,该类多边形在内部存在空洞。其查询方法与简单多边形的查询方法相同,但在要素创建的过程中,需要输入多边形内部顶点以创建内部空洞。

```
var polygon=L. geoJSON(
{
type: 'Polygon',
coordinates: [
[
[116.6418457,41.08886719],
```

```
        [118.86657715,41.01196289],
        [119.1027832,38.3203125],
        [115.00014648,37.74328613],
        [116.6418457,41.08886719]
    ],[
        [115.36508793228916,38.312993059223494],
        [118.30942386978916,38.312993059223494],
        [117.30942386978916,40.257328996723494],
        [115.36508793228916,38.312993059223494]
    ]]
},
{style:{color:'red',fillOpacity:0.1}}
).addTo(map);
```

运行效果如图6-6所示。

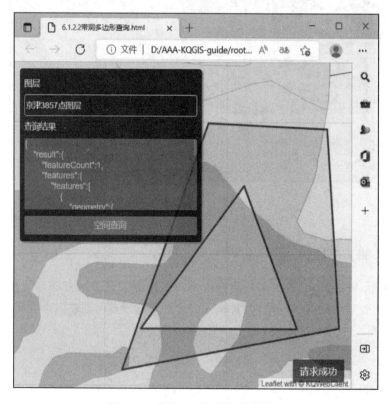

图6-6 带洞多边形查询效果预览

6.1.3 属性查询

在上传要素图层的同时,也会将要素的属性一同上传,如需按照特殊条件对要素进行筛选,则需使用属性查询,主要步骤如下。

第一步,完成地图初始化设置及地图导入。

第二步,设置出图地址,以出图地址作为参数,使用 L.KqGIS.tileMapLayer 导入图层,格式为 png,输出图层为 1~7,统一使用 addTo 方法将图层加入地图。

```
let jingjinUrl =
service_ip + (isMicroService ? '/jingjin3857 - proxy' : '') + '/kqgis/rest/services/jingjin3857/map';
let dataUrl =
service_ip + (isMicroService ? '/jingjin3857 - proxy' : '') + '/kqgis/rest/services/jingjin3857/data';
//添加图层
L.KqGIS.tileMapLayer(jingjinUrl,{
layerIds:[1,2,3,4,5,6,7],
format:'image/png'
}).addTo(map);
```

第三步,初始化 kendo 容器,为 query 容器绑定"kendoButton"按钮,当按钮被按下时,执行函数。首先使用 KqGIS.SQLFilterParams 创建筛选条件[where:"ADMINNAME='北京'"],然后使用 KqGIS.Data.GetFeaturesBySQLParams 设置查询参数对象,该类参数如表 6-3 所示。设置完参数列表后。使用 getFeaturesBySQL 方法,传入服务地址及查询参数 params,执行要素属性查询。

表 6-3　GetFeaturesBySQLParams 参数介绍

Name	Type	Default	参数定义
datasourceName	string		可选,数据源名称
datasourceID	string		可选,数据源 ID。与 datasourceName 参数功能相同,用于确定数据源
connInfo	KqGIS.Data.ConnectionInfoParams		可选,数据源连接信息,动态打开数据源。功能与其他 datasourceID 和 datasourceName 参数类似,优先级高于其他。该参数用于全局通用数据服务,根据临时连接信息动态打开数据源
datasetName	string		可选,数据集名称。需要与 datasourceName 或 datasourceID 参数组合使用,确定唯一数据集

续表 6-3

Name	Type	Default	参数定义
datasetID	string		可选,数据集 ID。可替代 datasourceName、datasourceID、datasetName 的组合,独自确定唯一数据集。使用 options.connInfo 参数时,本参数无效,请使用 options.datasetName
layerName	string		可选图层名称,地图服务下生效。可替代 datasourceName、datasourceID、datasetName、datasetID 参数,独自确定唯一数据集
layerId	string		可选,图层 ID,地图服务下生效。与 layerName 参数功能相同,二选一,优先级高于 layerName
attributeFilter	KqGIS.SQLFilterParams		属性过滤条件对象
fromIndex	number	0	可选,查询结果的最小索引号,整数。默认值是 0,如果该值大于查询结果的最大索引号,则查询结果为空
toIndex	number		可选,查询结果的最大索引号,整数。如果该值大于查询结果的最大索引号,则以查询结果的最大索引号为终止索引号
returnContent	boolean	false	可选,是否返回要素信息(属性和图形信息),默认为 false,只返回要素的 id。true 时,返回要素信息
hasGeometry	boolean	false	可选,是否返回要素图形信息,默认为 false,不返回要素图形信息;true 时,返回要素图形信息。在 returnContent 为 true、sync 为 false 时,该参数有效
outSRS	KqGIS.ServiceSRS		可选,结果数据集的空间参考
asyn	boolean	false	可选,false 表示直接返回查询结果,即元素类型为 Feature 的数组;true 表示异步执行查询,返回创建的 featureResult 资源的 name。默认 true

```
setTimeout(function () {
//初始化 kendo 容器
$ ('#query')
.kendoButton({
click: function () {
let filter=new KqGIS.SQLFilterParams({ where: "ADMINNAME='北京'" });
let params=new KqGIS.Data.GetFeaturesBySQLParams({
layerId: 2,
```

```
attributeFilter: filter,
// fromIndex: 1,
// toIndex: 2,
returnContent: true,
hasGeometry: true
});
L.KqGIS.dataFeatureService(dataUrl).getFeaturesBySQL(params, onsuccess, onfailed);
}
})
.data('kendoButton');
```

第四步,设置 kendoUI 的信息框,使用 disableScrollPropagation 终止事件派发,阻止设置的容器被分派到其他 Document 节点。

第五步,创建 onsuccess 查询成功回调函数,若查询成功,则使用 formatJSON 方式将 response 的结果 result 转换为 JSON 格式并将其值赋给 result 容器的 val 属性值。然后调用 L.geoJSON 将 result.features 加入地图。并在信息框中提示"请求成功",否则提示"请求失败"。

```
var onsuccess = function (response) {
$('#result').val(formatJSON(response.result));
L.geoJSON(response.result.result.features).addTo(map);
notification.show({ message: '请求成功' },'info');
};
var onfailed = function () {
notification.show({ message: '请求失败' },'info');
};
});
}
</script>
```

运行效果如图 6-7 所示。

6.1.4 几何+属性组合查询

几何+属性组合查询可用于在特定范围内查找指定属性值的要素,实现过程是在属性查询的基础上增加几何条件,具体实现代码如下。

使用 L.geoJSON 创建空间要素对象,设置其类型为多边形 polygon,设置多边形坐标点及要素颜色、透明度,使用 addTo 方法将其加入地图。

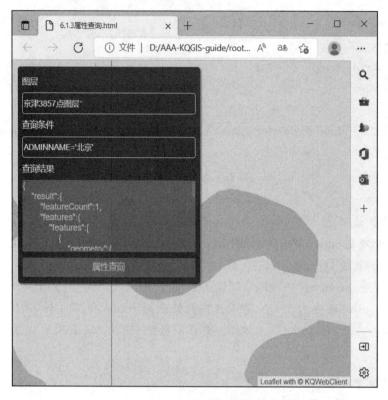

图 6-7 属性查询效果预览

```
var polygon=L.geoJSON(
{
type: 'Polygon',
coordinates: [
[
[115.36508793228916,38.312993059223494],
[118.30942386978916,38.312993059223494],
[118.30942386978916,41.257328996723494],
[115.36508793228916,41.257328996723494],
[115.36508793228916,38.312993059223494]
]
]
},
{ style: { color: 'red',fillOpacity: 0.03 } }
).addTo(map);
```

参考 6.1.3 设置 kendo 容器，使用 KqGIS.ServiceSRS 设置几何参考系，类型为 EPSG，值为 4326。在查询条件中，除了设置属性过滤条件外，还需在 filter 中加入 geometry、geosrs 以及空间关系类型，选择 INTERSECT 相交。

```
setTimeout(function () {
let attributeFilter = new KqGIS.SQLFilterParams({ where: " ADMINNAME = '北京'"});
let filter=new KqGIS.FilterParam({
attributeFilter: attributeFilter,
geometry: L.Util.toKqGISGeometry(polygon),
geoSRS: geoSRS,
SpatialRel: KqGIS.SpatialRel.INTERSECT
});
```

运行效果如图 6-8 所示。

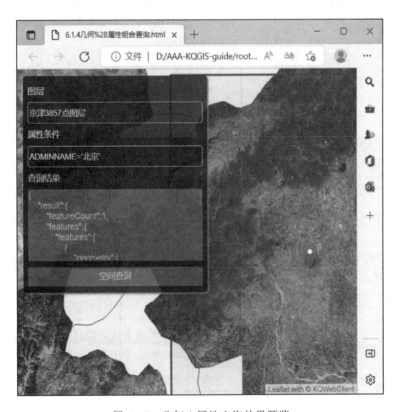

图 6-8　几何＋属性查询效果预览

6.1.5 根据要素唯一值查询

有时,范围查找的数据量过于冗杂,需要根据要素唯一值查找特定要素。要素唯一值查询本质上也是属性查询,故其他操作步骤与 6.1.3 均相同,其查询条件核心代码如下。

初始化 kendo 容器后,使用 KqGIS. Data. GetFeaturesByIDsParams 类创建查询参数列表,该类的具体参数如表 6-4 所示。然后调用 getFeaturesByIDs 方法进行查询,筛选指定 ID 的要素。

表 6-4 GetFeaturesByIDsParams 参数介绍

Name	Type	Default	参数定义
datasourceName	string		可选,数据源名称
datasourceID	string		可选,数据源 ID。与 datasourceName 参数功能相同,用于确定数据源
connInfo	KqGIS. Data. ConnectionInfoParams		可选,数据源连接信息,动态打开数据源。功能与其他 datasourceID 和 datasourceName 参数类似,优先级高于其他。该参数用于全局通用数据服务,根据临时连接信息动态打开数据源
datasetName	string		可选,数据集名称。需要与 datasourceName 或 datasourceID 参数组合使用,确定唯一数据集
datasetID	string		可选,数据集 ID。可替代 datasourceName、datasourceID、datasetName 的组合,独自确定唯一数据集。使用 options. connInfo 参数时,本参数无效,请使用 options. datasetName
layerName	string		可选,图层名称,地图服务下生效。可替代 datasourceName、datasourceID、datasetName、datasetID 参数,独自确定唯一数据集
layerId	string		可选,图层 ID,地图服务下生效。与 layerName 参数功能相同,二选一,优先级高于 layerName
ids	Array.<number>		需要过滤的要素 ID 数组,整数
hasGeometry	boolean	false	可选,是否返回要素图形信息,默认为 false,不返回要素图形信息,true 时,返回要素图形信息
outSRS	KqGIS. ServiceSRS		可选,结果数据集的空间参考

运行效果如图 6-9 所示。

图 6-9　根据要素唯一值查询

6.2　编辑要素

在 GIS 系统的使用过程中,经常需要对现有要素进行编辑,通常可分为增、删、改、查 4 个基础应用。在调取具体的增删改查接口时,需首先进行如下操作。

第一步,进行地图初始化和容器初始化,使用 addTo 方法将创建好的容器和天地图矢量图层及矢量注记图层加入地图中。

第二步,初始化数组 insertData 为空,seldata 列表为空,变量 indexDraw 初始值为 0,创建变量 drawRectPointHandler,创建 griddata 数组,存放 NAME、TYPE、KCMJ 三个列表信息。设置地图服务出图地址及数据地址。

```
let insertData=[];
let indexDraw=0; //
var drawRectPointHandler;
letgriddata=[{
"field": "NAME",
"value": ""
},{
"field": "TYPE",
"value": ""
},{
"field": "KCMJ",
```

```
"value": 1
}];
let seldata={};  //当前选择的对象
let mapurl=service_ip + (isMicroService ? '/edit4490-proxy' : '') + '/kqgis/rest/services/edit4490/map';
letdataUrl=service_ip + (isMicroService ? '/edit4490-proxy' : '') + '/kqgis/rest/services/edit4490/data';
```

第三步,设置默认图层 id 为 3,以 mapurl 和 layerid 为参数使用 L. KqGIS. tileMapLayer 创建图层对象,使用 addTo 方法将图层加入地图。设置图层属性信息,其中 typegroup 为 poygon 多边形,modelgroup 为 add 添加。然后初始化剩余变量 collections、datagrid、grid 等。创建 collect 数组,存放一个列表,文本域为"面图层",值域为"3"。使用 L. Draw. Marker 类在 map 中创建点绘制对象。同样地,使用 L. Draw. Polyline 和 L. Draw. Polygon 创建线、面绘制对象。

```
let layerid=3;
//添加图层
var maplayer=L. KqGIS. tileMapLayer(mapurl,{ layerIds: [layerid] }). addTo(map);
let typegroup = 'polygon', modelgroup = 'add', collections, datagrid, grid, drawlist = [], drawlayer = null;;
let collect=[{ text: '面图层', value: '3' }];
var pointDrawer=new L. Draw. Marker(map);
var lineDrawer=new L. Draw. Polyline(map);
var polygonDrawer=new L. Draw. Polygon(map);
```

第四步,使用 setTimeout 自动调用函数,为 typegroup 容器绑定 kendoRadioGroup。首先,设置 items 数组,存放 3 个列表,列表的 label 分别为点数据、线数据、面数据,值分别为 point、line、polygon。设置 layout 展示形式为 horizontal 水平放置。初始值为 polygon。

```
setTimeout(function () {
//初始化 kendo 容器
$("#typegroup").kendoRadioGroup({
items: [{
label: "点数据",
value: "point"
},{
label: "线数据",
value: "line"
```

```
},{
label："面数据",
value："polygon"
}],
layout："horizontal",
value："polygon",
```

创建选取事件，当某个 items 被选中时，首先用 jQuery 查询该 item 下的 target[0] 对象，使用 val()方法将其转为数值，使用 typegroup 承接。然后调用 cancelall 函数(后续步骤具体介绍)，如果此时 typegroup 的值为 point，则设置 layerid 为 1，collect 数组文本域为点图层，值域为 1。然后把 collect 设置为 collections 的数据源，再使用 select 方法选择 collections 的所有文本。同样地，分别设置线图层和面图层的选项。最后使用 map.removeLayer 清空原本图层，并使用 L.KqGIS.tileMapLayer 传入当前所选 items 的 layerid，创建新的图层。

```
select：function (e) {
typegroup=$(e.target[0]).val();
cancelall();
if (typegroup === "point") {
layerid=1;
collect=[{ text：'点图层',value：'1' }];
collections.dataSource.data(collect);
collections.select(0);
}
if (typegroup === "line") {
layerid=2;
collect=[{text：'线图层',value：'2' }];
collections.dataSource.data(collect);
collections.select(0);
}
if (typegroup === "polygon") {
layerid=3;
collect=[{ text：'面图层',value：'3' }];
collections.dataSource.data(collect);
collections.select(0);
}
map.removeLayer(maplayer);
maplayer=L.KqGIS.tileMapLayer(mapurl,{ layerIds：[layerid] }).addTo(map);
```

参考上一步,为 modelgroup 绑定 kendoRadioGroup,其中 item 的 label 分别为添加、修改、删除。Value 分别为 add、edit、del。Layout 设置为 horizontal 水平,初始值设为 add。创建选择事件,先调用 cancelall 函数,然后使用 jQuery 查询该 item 下的 target[0]并用 sel 承接,如果 sel 的值为 add,则设置 editbtn 的值为 true,其余选项的值为 false。同理,设置其他两种情况的 sel 值。最后设置 modelgroup 的默认值为 sel。

```javascript
$("#modelgroup").kendoRadioGroup({
items:[{
label:"添加",
value:"add"
},{
label:"修改",
value:"edit"
},{
label:"删除",
value:"del"
}],
layout:"horizontal",
value:"add",
select: function (e) {
cancelall();
let sel=$(e.target[0]).val();
if (sel==="add") {
selbtn.enable(false);
delbtn.enable(false);
editbtn.enable(true);
}
if (sel==="edit") {
selbtn.enable(true);
editbtn.enable(false);
delbtn.enable(false);
}
if (sel==="del") {
selbtn.enable(true);
delbtn.enable(false);
editbtn.enable(false);
}
```

```
modelgroup=sel;
}
});
```

第五步,为 collections 绑定"kendoDropDownList"下拉列表,设置数据源为 collect,文本域为 text,值域为 value,fillMode 为 none 不自动填充。

```
collections= $ ('#collections').kendoDropDownList({
dataSource: collect,
dataTextField: 'text',
dataValueField: 'value',
fillMode: "none"
}).data('kendoDropDownList');
```

为 datagrid 容器绑定 kendoGrid,设置高度和缩放标签,在 columns 中添加标题和 field 信息。添加 change 函数,当该容器的值变化时,使用 this 指针访问 dataItem 的 this.select 对象。如果访问的对象 properties 存在,则使用 JSON.parse 将 properties 对象转为 JSON 格式,并用 properties 变量储存,随后创建 griddata 数组,储存 KID、NAME、TYPE、KCMJ 四个列表,其中,列表的值为 properties 对应的子属性,layid 即为 selected 的 layid。

```
datagrid= $("#datagrid").kendoGrid({
height: 350,
scrollable: false,
selectable: true,
columns: [
{
field: "name", title: "图形列表"
}
],
change: function () {
var selected=this.dataItem(this.select());
if (selected.properties) {
let properties=JSON.parse(selected.properties);
griddata=[{
"field": "KID",
"value": properties["KID"],
"layid": selected.layid
},{
"field": "NAME",
```

```
"value": properties["NAME"],
"layid": selected.layid
},{
"field": "TYPE",
"value": properties["TYPE"],
"layid": selected.layid
},{
"field": "KCMJ",
"value": Number(properties["KCMJ"]),
"layid": selected.layid
}];
```

使用 removeGeometryByIdCustom 移除 id 为 result 的几何要素。用 JSON.parse 函数将 selected.geometry 转为 JSON 数组，然后用 locationDraw 在数组限定的位置绘制 result 要素。创建 seldata 变量，其初始值为 selected。如果 modelgroup 的值为 add 添加，则传入参数 t 调用 drawlist.map 函数判断当前地图是否有 t.layer，如果有，将其移除。

```
removeGeometryByIdCustom('result');
locationDraw(JSON.parse(selected.geometry),"result");
seldata=selected;
if (modelgroup === "add") {
drawlist.map(t => {
if (map.hasLayer(t.layer)) {
map.removeLayer(t.layer);
}
});
```

第六步，使用 drawlist.find 函数，传入参数 t(指代当前操作的对象)，若 t 的 layid 恒等于 selected 的 layid，则使用 removeLayer 移除 drawlayer 图层，再用 addLayer 方法添加 draw 对象创建的图层。将上一步中创建的 seldata 变量的值改为 selected。然后，调用 griddata 的 map 方法，传入参数 t，将 t 的 layid 属性值修改为 selected 的 layid。使用 setOptions 方法设置 grid 的 dataSource 数据源为 griddata。

```
let draw=drawlist.find(t => t.layid === selected.layid);
map.removeLayer(drawlayer);
map.addLayer(draw.layer);
}
seldata=selected;
```

```
griddata.map(t => {
t.layid=selected.layid;
});
grid.setOptions({
dataSource: griddata
});
},
}).data('kendoGrid');
```

第七步，为 grid 容器绑定 kendoGrid，设置高度和展开禁用，根据 field 值确定 column 展示的样式。

```
grid=$("#grid").kendoGrid({
height: 350,
scrollable: false,
columns: [
{ field: "field",title: "字段名",width: "90px" },
{
field: "value",title: "字段值",width: "120px",template: "# if (field == 'KCMJ') {
#<input type='number' name='txtValue' style='width:90px' value=#=value#></input>#  } else if (field == 'KID') {# #=value#  #}else{#<input type='text' style='width:90px' name='txtValue' value='#=value#'></input># } #"
}
]
}).data('kendoGrid');
```

为 grid 绑定事件，当监听到 txtValue 模糊输入时，设置 jQuery 对象 tr 的值为 this 指针指向的对象父节点的 tr 值。设置 jQuery 对象 uid 的值为 tr 对象的属性 data-uid。使用 find 函数查询 grid 的 tbody 的 tr 对象，创建 row 变量承接查询结果。将 row 显示在 grid 的 kendoGrid 的数据列表中。

```
$("#grid").on('blur','INPUT[name=txtValue]',function (e) {
$tr=$(this).parents('tr');
$uid=$tr.attr("data-uid")
var row=$('#grid').find("tbody->tr[data-uid=" + $uid + "]");
var model=$('#grid').data("kendoGrid").dataItem(row);
```

遍历 insertData 数组中每一个元素，如果元素的 layid 值等于 model 的 layid 值，则使用 JSON.parse 方法将元素的 properties 对象转化为 JSON，创建 propertie 变量承接。如果

propertie 为空,则将 propertie 重写为空列表。使用 jQuery 的 $ 查询方法取 this 指针指向的对象的 val 值,将该值赋给 propertie 列表中的 model.field 属性。赋值完成后,使用 JSON.stringify 将 propertie 对象转为字符串。然后将字符串赋值给 insertData 数组中的元素的 properties 属性。

```
for (let i=0; i < insertData.length; i++) {
if (insertData[i].layid === model.layid) {
let propertie = JSON.parse(insertData[i].properties);
if (propertie == null) {
propertie={};
}
propertie[model.field]=$(this).val();
insertData[i].properties=JSON.stringify(propertie);
}
}
});
```

为 save、cancel 容器添加 kendoButton。为 selbtn 绑定鼠标事件,当监听到鼠标点击时,使用 document.getElementById 查询容器 id 为 map 的对象,修改其样式 style.cursor 为 crosshair。调用 startDrawRectPoint 函数,传入参数 e 指代当前对象,当传入状态为 true 时,用变量 JSONGeometry 承接 e,然后以 JSONGeometry 为参数调用 infoquery 函数(下一步介绍)。

```
$("#save").kendoButton({
themeColor: "primary"
});
$('#cancel').kendoButton().data('kendoButton');
$("#selbtn").on('click',function () {
document.getElementById('map').style.cursor='crosshair';
startDrawRectPoint(e => {
var JSONGeometry=e;
infoquery(JSONGeometry);
},true);
});
```

第八步,创建 infoquery 函数,使用 removeGeometryByIdCustom 方法移除 id 为 result 的几何要素,使用 document.getElementById 查询容器 id 为 map 的对象,设置其样式 style.cursor 为 default。使用 KqGIS.ServiceSRS 创建坐标参考系对象 geoSRS,设置其 type 为 EPSG,值为 4326,使用 KqGIS.FilterParams 创建筛选条件,其中,图形要素为 JSONGeometry,坐标系为 geoSRS,查询条件为 SpatialRel.INTERSECT 相交查询。

使用 KqGIS.Data.GetFeaturesByGeometryParams 创建查询参数列表对象 params，该类参数如表 6-5 所示，根据列表信息填写参数信息。

表 6-5　GetFeaturesByGeometryParams 参数介绍

Name	Type	Default	参数定义
datasourceName	string		可选，数据源名称
datasourceID	string		可选，数据源 ID。与 datasourceName 参数功能相同，用于确定数据源
connInfo	KqGIS.Data.ConnectionInfoParams		可选，数据源连接信息，动态打开数据源。功能与其他 datasourceID 和 datasourceName 参数类似，优先级高于其他。该参数用于全局通用数据服务，根据临时连接信息动态打开数据源
datasetName	string		可选，数据集名称。需要与 datasourceName 或 datasourceID 参数组合使用，确定唯一数据集
datasetID	string		可选，数据集 ID。可替代 datasourceName、datasourceID、datasetName 的组合，独自确定唯一数据集。使用 options.connInfo 参数时，本参数无效，请使用 options.datasetName
layerName	string		可选，图层名称，地图服务下生效。可替代 datasourceName、datasourceID、datasetName、datasetID 参数，独自确定唯一数据集
layerId	string		可选，图层 ID，地图服务下生效。与 layerName 参数功能相同，二选一，优先级高于 layerName
filter	KqGIS.FilterParams		查询过滤条件对象
fromIndex	number	0	可选，查询结果的最小索引号，整数。默认值是 0，如果该值大于查询结果的最大索引号，则查询结果为空
toIndex	number		可选，查询结果的最大索引号，整数。如果该值大于查询结果的最大索引号，则以查询结果的最大索引号为终止索引号
returnContent	boolean	false	可选，是否返回要素信息（属性和图形信息），默认为 false，只返回要素的 id。true 时，返回要素信息
hasGeometry	boolean	false	可选，是否返回要素图形信息，默认为 false，不返回要素图形信息，true 时，返回要素图形信息。在 returnContent 为 true，sync 为 false 时，该参数有效
outSRS	KqGIS.ServiceSRS		可选，结果数据集的空间参考
asyn	boolean	false	可选，false 表示直接返回查询结果，即元素类型为 Feature 的数组；true 表示异步执行查询，返回创建的 featureResult 资源的 name。默认不传时为 true

```
//拉框查询
function infoquery(JSONGeometry) {
removeGeometryByIdCustom('result');
document.getElementById('map').style.cursor='default';
let geoSRS=newwindow.KqGIS.ServiceSRS({
type：KqGIS.ProjectSystemType.EPSG,
value：'4326'
});
let filter=new window.KqGIS.FilterParams({
geometry：JSONGeometry,
geoSRS：geoSRS,
SpatialRel：window.KqGIS.SpatialRel.INTERSECT
});
let params=new window.KqGIS.Data.GetFeaturesByGeometryParams({
layerId：layerid,
filter：filter,
returnContent：true,
hasGeometry：true,
});
```

第九步，传入 dataUrl 地图服务出图地址、params 等参数，调用 dataFeatureService 方法进行查询，返回结果 response。使用变量 res 存放返回结果的要素 response.result.features，如果 response.result.resul 的要素个数大于 0，则设置 insertData 为空值，设置 indexDraw 值为 0。然后使用 forEach 遍历每个要素。传入 val 作为参数，设置 layerid 为"insert+当前序号"，使用 Math.random 随机数取 36 位，用 tostring 方法写为字符串，然后取第 3~12 位作为 id。使用 id layid 初始化 feature，用 val 的对应属性设置 name、properties、geometry 等信息。

```
    L.KqGIS.dataFeatureService(dataUrl).getFeaturesByGeometry(params,function(response){
    let res=response.result.features;
    if (response.result.result.featureCount＞0) {
    insertData=[ ];
    indexDraw=0;
    res.forEach(function (val) {
    let layerid='insert' + indexDraw++;
    let id=Math.random().toString(36).substr(3,12);
    let feature={
    id：id,
    layid：layerid,
```

```
name:'地块' + indexDraw,
properties: JSON.stringify(val.properties),
geometry: JSON.stringify(val.geometry)
};
```

使用 push 方法将 feature 压入 insertData 数组。以 val.geometry 为参数,调用 locationDraw 进行绘制,将结果显示在 result 中。然后使用 setOptions 方法将 datagrid 的数据源设置为 insertData,将 grid 的数据源设置为空值。

```
insertData.push(feature);
locationDraw(val.geometry,"result");
});
datagrid.setOptions({
dataSource: insertData
});
grid.setOptions({
dataSource: []
})
}
});
}
```

第十步,为 map 添加事件,当监听到 draw 对象被创建时,如果 drawlayer 存在,则将其移除。然后修改 drawlayer 的值为 e 指向的对象的 layer,并使用 addTo 方法加入地图。

```
map.on('draw:created',function (e) {
if (drawlayer) {
map.removeLayer(drawlayer);
}
drawlayer=e.layer;
drawlayer.addTo(map);
```

设置 delbtn 编辑按钮的启用状态为 true。如果此时 modelgroup 为 add,将 drawlayer 作为参数传入,调用 addInsertData 函数。使用 setOptions 方法将 datagrid 的数据源设置为 insertData。将 drawlayer 作为参数传入,调用 updateInsertData 函数。

```
delbtn.enable(true);
if (modelgroup === "add") {
addInsertData(drawlayer);
datagrid.setOptions({
dataSource: insertData
});
```

```
} else {
updateInsertData(drawlayer);
}
});
```

为 editbtn 绑定点击事件,当按钮被按下后,用 if 语先判断 typegroup 的类型,然后调用相应的 drawer。

```
$("#editbtn").on('click',function () {
if (typegroup === "point") {
pointDrawer.enable();
}
if (typegroup === "line") {
lineDrawer.enable();
}
if(typegroup === "polygon") {
polygonDrawer.enable();
}
});
```

为 delbtn 绑定点击事件,当按钮被按下后,如果 drawlayer 存在,则将其移除。用 if 语句判断当前 modelgroup 的类型,以 add 为例,如果当前 seldata 的 layid 存在。则遍历 insertData,取数据中与 seldata 元素的 layid 相同的元素,将其删除后返回。若不存在 layid,则将 insertData 设置为空值。然后使用 setOptions 方法将 datagrid 的数据源设置为 insertData,将 grid 的数据源设置为空值。

```
$("#delbtn").on('click',function () {
if (drawlayer) {
map.removeLayer(drawlayer);
}
if (modelgroup === "add") {
if (seldata.layid) {
for (let i=0; i < insertData.length; i++) {
if (insertData[i].layid === seldata.layid){
insertData.splice(i,1);
break;
}
}
} else {
```

```
insertData=[];
}
datagrid.setOptions({
dataSource：insertData
});
grid.setOptions({
dataSource：[]
});
}
```

第十一步，创建 updateInsertData 函数，传入 layer 作为参数，当该函数调用时，使用 toGeoJSON 方法将 layer 转化为 geoJSON 数据，用 features 承接。使用变量 geo 储存 features 的属性。使用 JSON.parse 方法将 seldata.properties 对象转为 JSON，使用变量 properties 承接。用 for 循环遍历 insertData 数组的每一个元素，用 JSON.parse 把元素的 properties 属性转为 JSON，用 prop 承接。用 if 语句判断 prop 的 KID 属性是否与 properties 数组中的一致。若一致，则使用 JSON.stringify 将 geo 对象转为字符串，将字符串的值赋给 insertData 数组元素的 geometry 属性，完成画图的更新。

```
//更新画图
function updateInsertData(layer) {
let features=layer.toGeoJSON();
let geo=features.geometry;
let properties=JSON.parse(seldata.properties);
for (let i=0; i < insertData.length; i++) {
let prop=JSON.parse(insertData[i].properties);
if (prop.KID === properties["KID"]) {
insertData[i].geometry=JSON.stringify(geo);
break;
}
}
}
```

第十二步，创建 featurelist 数组，初始化其值为空。For 循环遍历 propertielist 数中的每一个元素，以该元素的 geometry 和 properties 属性，使用 KqGIS.Feature 创建要素对象 feature。使用 push 方法，将 feature 压入 featurelist 数组。

```
//构造 Feature 或 FeatureCollection
let featurelist=[];
for (let y=0; y < propertielist.length; y++) {
```

```
    let feature=new KqGIS. Feature(propertielist[y]. geometry,propertielist[y]. proper-
ties);
    featurelist. push(feature);
}
```

以 featurelist 为参数,使用 KqGIS. FeatureCollection 创建要素数据集对象 featurecollection。用 KqGIS. Format. GeoJSON 类创建 geoJSONFormat 对象。使用 geoJSONFormat 的 write 方法将几何对象 featurelcollection 写成 GeoJSON 对象,用变量 featureGeoJSON 承接。使用 KqGIS. ServiceSRS 创建坐标参考系对象 geoSRS ,设置其 type 为 EPSG,值为 4326。使用 KqGIS. Data. GetFeaturesByGeometryParams 创建查询参数列表对象 params 传入 dataUrl 地图服务出图地址、params 等参数,调用 dataFeatureService 方法进行查询,设置查询回调函数。

```
let featurelcollection=new KqGIS. FeatureCollection(featurelist);
let geoJSONFormat=new KqGIS. Format. GeoJSON();
let featureGeoJSON =geoJSONFormat. write(featurelcollection);
let geoSRS=new KqGIS. ServiceSRS({
type：KqGIS. ProjectSystemType. EPSG,
value：'4326'
});
let params=new KqGIS. Data. ImportFeaturesParams({
layerId：layerid,
features：featureGeoJSON,
geoSRS：geoSRS
});
L. KqGIS. dataFeatureService(dataUrl). importFeatures(params,create_onsuccess,create_onfailed);
}
```

第十三步,创建 stopDrawRectPoint 函数关闭拉框查询的状态,如果 drawRectPointHandler 对象不为 undefined,则关闭所有鼠标事件。然后以 feature、locationDrawStyle 为参数使用 L. geoJSON 创建要素 tempLayer。如果当前 layerGroups[idCustom]对象未被定义,则使用 L. featureGroup 创建对象并加入地图。然后将 tempLayer 分别加入 layerGroups[idCustom]和 map 中。最后调整地图显示范围至合适位置。

```
function stopDrawRectPoint() {
if (drawRectPointHandler ! == undefined) {
map. off('mousemove',drawRectPointHandler. onMove);
```

```
map.off('mouseup',drawRectPointHandler.onUp);
map.off('mousedown',drawRectPointHandler.onDown);
}
}
let tempLayer=L.geoJSON(feature,locationDrawStyle);
if (layerGroups[idCustom] === undefined) {
layerGroups[idCustom]=L.featureGroup().addTo(map);
}
layerGroups[idCustom].addLayer(tempLayer);
map.addLayer(tempLayer)
map.fitBounds(tempLayer.getBounds(),{
paddingTopLeft：[200,200],
paddingBottomRight：[200,200]
});
```

创建 removeGeometryByIdCustom 移除指定 id 的几何信息的函数。

```
function removeGeometryByIdCustom(idCustom) {
if (layerGroups[idCustom] !== undefined) {
layerGroups[idCustom].clearLayers();
map.removeLayer(layerGroups[idCustom]);
delete layerGroups[idCustom];
}
}
```

最后一步,创建数据清空函数 cleardata,当该函数调用时,设置 insertData 数组的值为空值。使用 setOptions 方法设置 datagrid 和 grid 的数据源为空值。

```
//清除数据
function cleardata() {
insertData=[];
datagrid.setOptions({
dataSource：[]
});
grid.setOptions({
dataSource：[]
});
}
}
```

创建清空页面函数 cancelall，当该函数被调用时，使用 cleardata 清空数据，使用 removeGeometryByIdCustom 清空结果显示。使用 document.getElementById 设置地图的样式为 default。若此时 modelgroup 的状态为 add，则启用 editbtn；否则，将其禁用。最后，依次移除点线面的绘画工具。

```
function cancelall() {
cleardata();
removeGeometryByIdCustom("result");
document.getElementById('map').style.cursor='default';
stopDrawRectPoint();
if (modelgroup === "add") {
editbtn.enable(true);
} else {
editbtn.enable(false);
}
pointDrawer.disable();
lineDrawer.disable();
polygonDrawer.disable();
}
</script>
```

至此，前期准备工作完成，主界面的运行效果如图 6-10 所示。

图 6-10　要素编辑主界面

6.2.1 添加要素

添加要素是 GIS 系统中的常用功能，一般由用户在网页端进行自定义绘制，系统端进行同步更新。在完成 6.2 的准备步骤后，还需进行如下操作。

第一步，创建 insert 函数，当该函数调用时，创建 create_onsuccess 成功回调函数，当调用成功时，先使用 removeGeometryByIdCustom 移除 ID 为 result 的几何要素，调用 cancelall 函数。在 notification 消息框中添加 info，其值为"要素入库成功"。

```
//入库
function insert() {
varcreate_onsuccess=function () {
removeGeometryByIdCustom("result");
cancelall();
notification.show({ message：'要素入库成功'},'info');
//拖曳地图来刷新要素
```

第二步，使用 setTimeout 方法等待 300ms 后自动调用函数，先用 map.removeLayer 移除当前 maplayer，再传入 mapurl 和当前 layerids 使用 L.KqGIS.tileMapLayer 创建新的图层。

```
setTimeout(function () {
// map.removeLayer(maplayer);
map.removeLayer(maplayer);
maplayer=L.KqGIS.tileMapLayer(mapurl,{ layerIds：[layerid] }).addTo(map);
},300);
};
```

第三步，创建 create_onfailed 失败回调函数，在 notification 消息框中添加 info，其值为"要素入库失败"。创建空数组 propertielist，对 insertData 数组中的值进行遍历，创建空列表 properties，使用 JSON.parse 将 insertData 当前对象的 properties 属性转为 JSON，用 prop 变量承接。然后对 prop 中的 key 进行遍历，若 key 的值为 KCMJ，则使用 Number 函数将 prop[key]转为数值然后赋值给 properties[key]，否则，直接将 prop[key]赋值给 properties[key]。然后使用 JSON.parse 把 insertData[i].geometry 属性转为 JSON，再用 L.Util.toKqGISGeometry 将 JSON 对象转为 KqGIS 几何要素。创建 props 数组存放 properties，把 props 用 push 方法压入 propertielist 数组。

```
var create_onfailed=function () {
notification.show({ message：'要素入库失败'},'info');
};
let propertielist=[];
for (var i=0; i < insertData.length; i++) {
let properties_={};
```

```
let prop=JSON.parse(insertData[i].properties);
for (let key in prop) {
if (key === "KCMJ") {
properties[key]=Number(prop[key]);
} else {
properties[key]=prop[key];
}
}
let geo = L.Util.toKqGISGeometry(JSON.parse(insertData[i].geometry));
let props={ properties: properties, geometry: geo };
propertielist.push(props);
}
```

第四步,创建绘图函数 startDrawRectPoint,传入两个参数,第一个参数 callbackFun 是回调方法,返回值是坐标,第二个参数 isMultiple 是执行多次,还是一次,默认值为 false。然后调用 stopDrawRectPoint 停止当前绘制。然后创建 drawRectPointHandler 对象,其 tmprect 临时矩形属性值为 null,latlngs 数组为空值。Ondown 函数被触发时,会禁用地图的拖拽功能。然后用 this_变量储存 drawRectPointHandler 对象,如果 this_的 tmprect 不为空值,则将 tmprect 移除。然后读取当前鼠标位置的经纬度,将其存入 this_.latlngs 数组的第一位。添加 mousemove 和 mouseup 事件监听。

```
function startDrawRectPoint(callbackFun,isMultiple) {
stopDrawRectPoint();
drawRectPointHandler={
tmprect: null,
latlngs: [],
onDown: function (e) {
map.dragging.disable();
let this_=drawRectPointHandler;
if (this_.tmprect !== null) {
this_.tmprect.remove();
}
//左上角坐标
this_.latlngs[0]=[e.latlng.lat,e.latlng.lng];
//开始绘制,监听鼠标移动事件
map.on('mousemove',this_.onMove);
map.on('mouseup',this_.onUp);
},
```

当 onMove 事件触发时，修改 this_ 的值为 drawRectPointHandler，然后读取当前鼠标位置的经纬度，将其存入 this_.latlngs 数组的第二位。如果 tmprect 不为空值，则将 tmprect 移除。以 this_.latlngs 给出的经纬度信息为参数，使用 L.rectangle 创建临时矩形 this_.tmprect，设置其样式，用 addTo 方法将其加入地图。继续监听 mouseup 事件。

```
onMove: function (e) {
let this_=drawRectPointHandler;
this_.latlngs[1]=[e.latlng.lat,e.latlng.lng];
//删除临时矩形
if (this_.tmprect!==null) {
this_.tmprect.remove();
}
//添加临时矩形
this_.tmprect=window.L.rectangle(this_.latlngs,{
color:'#3385ff',
weight:3,
dashArray:1
}).addTo(map);
map.on('mouseup',this_.onUp);
```

onUp 事件触发时，标志矩形绘制完成，移除临时矩形，并停止监听鼠标移动事件。修改 this_ 的值为 drawRectPointHandler，如果 tmprect 不为空值，则将 tmprect 移除。关闭对 mousemove 和 mouseup 的监听，启动地图的 dragging 功能。

```
onUp: function (e) {
let this_=drawRectPointHandler;
if (this_.tmprect!==null) {
this_.tmprect.remove();
}
map.off('mousemove',this_.onMove);
map.off('mouseup',this_.onUp);
map.dragging.enable();
```

读取当前鼠标位置的经纬度，将其存入 this_.latlngs 数组的第二位。令 p1 点（左上）的坐标为 this_.latlngs[0][1]，this_.latlngs[0][0]，p3 点（右下）坐标为当前鼠标位置的经纬度。如果两点经纬度之差的绝对值均小于 0.000000000000001，则 callbackFun 返回的坐标类型为点，坐标值为 p1 的坐标。否则，返回的坐标类型为多边形，坐标值为[p3[0],p1[1]]，p3,[p1[0],p3[1]],p1,[p3[0],p1[1]]。继续监听 mousedown 事件。

```
this_.latlngs[1] =[e.latlng.lat,e.latlng.lng];
let p1=[this_.latlngs[0][1],this_.latlngs[0][0]];
let p3=[e.latlng.lng,e.latlng.lat];
if ((Math.abs(p3[0] - p1[0]) < 0.000000000000001) && (Math.abs(p3[1] - p1[1]) < 0.000000000000001)) {
callbackFun({ 'type': 'Point','coordinates': p1 });
} else {
callbackFun({
'type': 'Polygon',
'coordinates': [[[p3[0],p1[1]],p3,[p1[0],p3[1]],p1,[p3[0],p1[1]]]]
});
}
}
};
map.on('mousedown',drawRectPointHandler.onDown);
}
```

运行效果如图 6-11 所示。

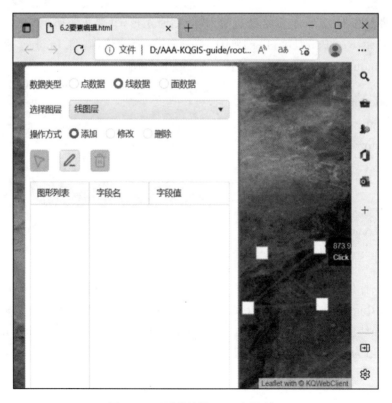

图 6-11 要素编辑——要素添加

6.2.2 修改要素

修改要素是针对已有要素的样式或者属性信息进行修改,一般由用户在网页端对指定要素进行自定义修改,具体步骤如下。

第一步,创建 update 函数,当该函数调用时,创建 update_onsuccess 成功回调函数,当调用成功时先在 notification 消息框中添加 info,其值为'要素更新成功'。

使用 removeGeometryByIdCustom 移除 id 为 result 的几何要素。

```
function update() {
var update_onsuccess=function () {
notification.show({ message:'要素更新成功' },'info');
removeGeometryByIdCustom("result");
//拖曳地图来刷新要素
```

第二步,使用 setTimeout 方法等待 300ms 后自动调用函数,先用 map.removeLayer 移除当前 maplayer,再传入 mapurl 和当前 layerids,使用 L.KqGIS.tileMapLayer 创建新的图层。

```
setTimeout(function () {
map.removeLayer(maplayer);
maplayer=L.KqGIS.tileMapLayer(mapurl,{ layerIds:[layerid] }).addTo(map);
map.panBy([1,0]);
},300);
};
```

第三步,创建 create_onfailed 失败回调函数,在 notification 消息框中添加 info,其值为"要素更新失败"。创建空数组 propertielist,对 insertData 数组中的值进行遍历,创建空列表 properties,使用 JSON.parse 将 insertData 当前对象的 properties 属性转为 JSON,用 prop 变量承接。然后对 prop 中的 key 进行遍历,若 key 的值为 KCMJ,则使用 Number 函数将 prop[key]转为数值然后赋值给 properties[key];否则,直接将 prop[key]赋值给 properties[key]。使用 JSON.parse 把 insertData[i].geometry 属性转为 JSON,再用 L.Util.toKqGISGeometry 将 JSON 对象转为 KqGIS 几何要素。创建 props 数组存放 properties,用 push 方法把 props 压入 propertielist 数组。

```
var update_onfailed=function () {
notification.show({ message:'要素更新失败' },'info');
};
let propertielist=[];
for (var i=0; i < insertData.length; i++) {
let properties={};
```

```
let prop=JSON.parse(insertData[i].properties);
for (let key in prop) {
if (key === "KCMJ") {
properties[key]=Number(prop[key]);
} else {
properties[key]=prop[key];
}
}
let geo=L.Util.toKqGISGeometry(JSON.parse(insertData[i].geometry));
let props={ properties: properties, geometry: geo };
propertielist.push(props);
}
let featurelist=[];
for (let y=0; y < propertielist.length; y++) {
let feature=new KqGIS.Feature(propertielist[y].geometry, propertielist[y].properties);
featurelist.push(feature);
}
```

第四步，以 featurelist 为参数，使用 KqGIS.FeatureCollection 创建要素数据集对象 featurelcollection。用 KqGIS.Format.GeoJSON 类创建 geoJSONFormat 对象。使用 geoJSONFormat 的 write 方法将几何对象 featurelcollection 写成 GeoJSON 对象，用变量 featureGeoJSON 承接。使用 KqGIS.Data.UpdateFeaturesParams 创建查询参数列表对象 params，该类参数如表 6-6 所示。

表 6-6　UpdateFeaturesParams 参数介绍

Name	Type	参数定义
datasourceName	string	可选，数据源名称
datasourceID	string	可选，数据源 ID。与 datasourceName 参数功能相同，用于确定数据源
connInfo	KqGIS.Data.ConnectionInfoParams	可选，数据源连接信息，动态打开数据源。功能与其他 datasourceID 和 datasourceName 参数类似，优先级高于其他。该参数用于全局通用数据服务，根据临时连接信息动态打开数据源
datasetName	string	可选，数据集名称。需要与 datasourceName 或 datasourceID 参数组合使用，确定唯一数据集

续表 6-6

Name	Type	参数定义
datasetID	string	可选,数据集 ID。可替代 datasourceName、datasourceID、datasetName 的组合,独自确定唯一数据集。使用 options.connInfo 参数时,本参数无效,请使用 options.datasetName
layerName	string	可选,图层名称,地图服务下生效。可替代 datasourceName、datasourceID、datasetName、datasetID 参数,独自确定唯一数据集
layerId	string	可选,图层 ID,地图服务下生效。与 layerName 参数功能相同,二选一,优先级高于 layerName
features	GeoJSONObject	Feature 或 FeatureCollection 类型的 GeoJSON 对象
geoSRS	KqGIS.ServiceSRS	可选,features 中的图形空间参考,值空或者无此项时为当前要素表的空间参考
filter	KqGIS.FilterParams | KqGIS.SQLFilterParams	可选,过滤条件对象。要求 features 中 geojoson 不带 geometry 字段
ids	Array.<number>	可选,需要过滤的要素 ID 数组,整数,属于 filter 的一种特例,优先级高于 filter 对象参数

传入 dataUrl 地图服务出图地址、params 等参数,调用 dataFeatureService 方法进行查询,设置查询回调函数,完成要素修改更新。

```
let featurelcollection=new KqGIS.FeatureCollection(featurelist);
let geoJSONFormat=new KqGIS.Format.GeoJSON();
let featureGeoJSON=geoJSONFormat.write(featurelcollection);
let params=new KqGIS.Data.UpdateFeaturesParams({
layerId: layerid,
features: featureGeoJSON
});
L.KqGIS.dataFeatureService(dataUrl).updateFeatures(params,update_onsuccess,update_onfailed);
```

运行效果如图 6-12 所示。

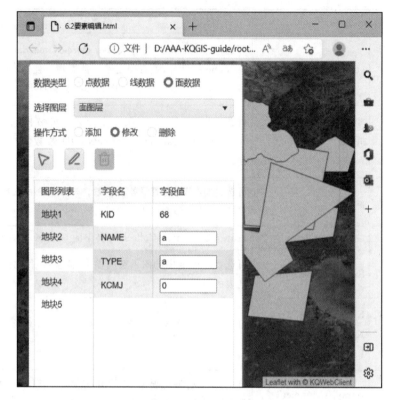

图 6-12　要素编辑——要素修改

6.2.3　删除要素

修改要素是针对已有的要素进行清除（包括属性信息），一般由用户在网页端对指定要素进行选中删除，系统端进行同步更新。具体步骤如下。

第一步，参考 6.2.1 将 props 压入 propertielist 数组，创建 filter 列表。如果 seldata. properties 存在，则使用 JSON. parse 将 seldata. properties 转为 JSON 格式，用 properties 变量承接。使用 KqGIS. SQLFilterParams 创建 SQL 查询条件 where：'KID=' + properties ["KID"]，使用 filter 承接；否则，创建空数组 kids，然后用 for 循环遍历 insertData 数组中每一个元素，使用 JSON. parse 方法将元素的 properties 转为 JSON 格式，用 prop 承接。取 prop 的 KID 属性，使用 push 方法将其压入 kids 数组中。

```
let filter={};
//单个删除
if (seldata. properties) {
let properties=JSON. parse(seldata. properties);
filter=newKqGIS. SQLFilterParams({ where：'KID=' + properties["KID"] });
} else {
```

```
let kids=[];for (let j=0; j < insertData.length; j++){
let prop=JSON.parse(insertData[j].properties);kids.push(prop.KID);
}
```

第二步,使用 KqGIS.SQLFilterParams 创建 SQL 查询条件 where:'KID in (' + kids.join(',') + ')',使用 filter 承接。使用 KqGIS.Data.DeleteFeaturesParams 创建查询参数列表对象 params,该类参数如表 6-7 所示。

表 6-7　DeleteFeaturesParams 参数介绍

Name	Type	参数定义
datasourceName	string	可选,数据源名称
datasourceID	string	可选,数据源 ID。与 datasourceName 参数功能相同,用于确定数据源
connInfo	KqGIS.Data.ConnectionInfoParams	可选,数据源连接信息,动态打开数据源。功能与其他 datasourceID 和 datasourceName 参数类似,优先级高于其他。该参数用于全局通用数据服务,根据临时连接信息动态打开数据源
datasetName	string	可选,数据集名称。需要与 datasourceName 或 datasourceID 参数组合使用,确定唯一数据集
datasetID	string	可选,数据集 ID。可替代 datasourceName、datasourceID、datasetName 的组合,独自确定唯一数据集。使用 options.connInfo 参数时,本参数无效,请使用 options.datasetName
layerName	string	可选,图层名称,地图服务下生效。可替代 datasourceName、datasourceID、datasetName、datasetID 参数,独自确定唯一数据集
layerId	string	可选,图层 ID,地图服务下生效。与 layerName 参数功能相同,二选一,优先级高于 layerName
filter	KqGIS.FilterParams | KqGIS.SQLFilterParams	可选,过滤条件对象
ids	Array.<number>	可选,需要过滤的要素 ID 数组,整数。属于 options.filter 的一种特例,优先级高。二选一

```
filter=new KqGIS.SQLFilterParams({ where: 'KID in (' +kids.join(',') + ')' });
let params=new KqGIS.Data.DeleteFeaturesParams({layerId: layerid,filter: filter
});
```

第三步,传入 dataUrl 地图服务出图地址、params 等参数,调用 dataFeatureService 方法进行查询,设置查询回调函数,完成要素修改删除。

L. KqGIS. dataFeatureService(dataUrl). deleteFeatures(params, delete_onsuccess, delete_onfailed);
　}});

运行效果如图 6-13 所示。

图 6-13　要素编辑——要素删除

7 KQGIS 几何服务

7.1 几何坐标投影转换

坐标投影转换用于解决 GIS 系统中坐标数据的统一问题,例如最基本的经纬度与平面直角坐标系互相转换。KQGIS 提供根据空间参考系信息投影和根据已有空间参考系 ID 投影的转换方法。

7.1.1 根据空间参考系信息投影

根据空间参考系进行坐标投影转换的步骤如下。

第一步,使用 L.map 创建地图对象 map 并设置地图投影中心、缩放层级。设置自定义缩放状态为 false。创建 gaodeTilelayer 图层,并调用 addTo 方法将矢量图层(vec)添加到地图中。

第二步,使用 L.Control 类自定义 GeometryProjection 控件,使用 DomUtil 类的 create 方法创建转换后的结果视窗容器(用法参考 https://www.wrld3d.com/wrld.js/latest/docs/leaflet/L.DomUtil),设置目标空间参考选项卡、坐标投影转换后结果栏、坐标投影转换按钮。随后定义 geometryProjection 函数,以 options 作为输入参数。调用 geometryProjection 方法创建 GeometryProjection 容器,使用 addTo 方法把新建的容器添加到地图的左上角。options 参数设置方法见第五步。

```
L.Control.GeometryProjection=L.Control.extend({
onAdd: function (map) {
var control_html =   L.DomUtil.create('div','geometry_projection');
var template=`<div id="maptools">
<div>
<span>目标空间参考</span>
<input value="EPSG:4326" id="outSRS" type="text">
</div>
<div>
<span>坐标投影转换后结果</span>
<textarea id="result" class="k-textbox" style="resize:none"></textarea>
```

```
</div>
<button id="switch">坐标投影转换</button>
</div>
`;
setTimeout(function () {
$(".geometry_projection").append(template);
});
return control_html;
},
});
L.control.geometryProjection=function(opts) {
return new L.Control.GeometryProjection(opts);
};
L.control.geometryProjection({position:'topleft'}).addTo(map);
```

第三步,创建 coordinates 数组存放 4 个端点的坐标。基于 L.polygon 类,创建 polygon 对象(Polygon 是继承自 Polyline 的一个类,它的生成函数是 L.polygon(<LatLng[]> latlngs, <Polyline options> options?)),将 coordinates 作为 latlng 参数传入生成函数,设置 style 属性,将矩形框颜色设置为绿色,线宽设为 1。随后使用 addTo 方法,将矩形框加入地图容器中。调用 fitBounds 方法,将矩形框设置为当前显示范围。

```
var coordinates=[[34.72,111.23],[34.69,111.23],[34.69,111.30],[34.72,111.30]];
var polygon=L.polygon(coordinates,{
style:{
"color":"green",
"weight":1
}
}).addTo(map);
map.fitBounds(polygon.getBounds());
```

第四步,选择 switch 容器,将其绑定为 kendoButton 按钮。同样地,将 outSRS 容器绑定为 kendoDropDownList,设置下拉列表选项为"EPSG:4326""EPSG:3857""EPSG:4490",将 notification 绑定为 kendoNotification,作为结果显示栏。设置 kendoNotification 的位置,自动隐藏时间设置为 2s,模板类型设置为系统信息,同时使用 kendoUI 信息的容器样式。完成绑定后,使用 disableScrollPropagation 终止事件派发,阻止第二步设置的容器被分派到其他 Document 节点。

第五步,定义 switch 函数,绑定鼠标单击事件。设置接口 options,包括端口基地址、待传数据(深色部分为待转换坐标点)、原始坐标系、outSRS 目标坐标系。以 options 为参数调用 Projection 坐标投影转换服务。具体参数信息如表 7-1 所示。

表 7 - 1   options 参数介绍

Name	Type	参数定义
url	string	kqservice 的 url
geometry	GeoJSONObject	需要进行投影转换的图形
geoSRS	String	传入图形的空间参考
outSRS	String	输出结果图形的空间参考

使用 queryAsync 设置回调函数，若成功，则将返回的 JSON 用 JSON.parse 解码，在结果视窗中显示，使用 flyToBounds 方法将地图缩放至第三步创建的 polygon 范围，使用 notification.show 弹出提示信息"请求成功"。

```
$("#switch").on("click",function () {
//接口 options
var options={};
options.url=service_ip + "/KQGis/rest/services";
options.data='{"type":"GeometryCollection","geometries":[{"type":"Polygon","coordinates":[[[111.23,34.72],[111.23,34.69],[111.30,34.69],[111.30,34.72]]]}]}';
options.geoSRS='EPSG:4326';
options.outSRS=$("#outSRS").val();
var projection=new L.kqmap.services.Projection(options);
projection.queryAsync(onsuccess,onfailed);
});
var onsuccess=function (response) {
$("#result").val(formatJSON(JSON.parse(response).result));
map.flyToBounds(polygon.getBounds());
notification.show({
message："请求成功"
},"info");
};
```

运行结果如图 7 - 1 所示。

7.1.2　根据已有空间参考系 ID 投影

根据已有空间参考系 ID 进行坐标投影转换的步骤如下。

第一步，参考 3.1 创建地图容器并初始化地图，设置地图投影中心、缩放层级，创建 gaodeTilelayer 图层，并调用 addTo 方法将矢量图层（vec）添加到地图中。

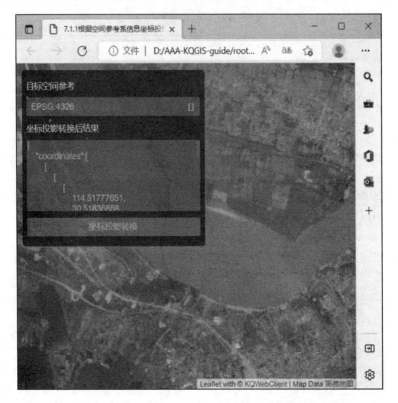

图 7-1　根据空间参考系信息投影

第二步，使用 L. Control. extend 类自定义 GeometryAttribute 控件，使用 DomUtil 类的 create 方法创建转换后的结果视窗容器。

第三步，为 analysis 控件绑定 kendobutton 按钮。为 epsg-code 容器绑定下拉列表，设置文本域为 code、值域为 index（选项序号），筛选条件为 contains。为 select 绑定鼠标事件，当鼠标点击"选项调用"后，令 code 值等于所选 dataItem 的 code 值。用 get 方法取到 code 的 name 及 wkt 值并分别赋值给 proj-name 和 result 变量。在 virtual 属性中设置 itemHeight 为 26。在外部设置 analysis 控件的 height 为 20。

```
setTimeout(function() {
//初始化 kendo 容器
$('#analysis').kendoButton().data('kendoButton');
let epsgList = $('#epsg-code').kendoDropDownList({
dataTextField: 'code',
dataValueField: 'index',
filter: 'contains',
select: function(e) {
let code = e.dataItem.code;
```

```
    $('#proj-name').val(coordinateSystemMap.get(code).name);
    $('#result').val(coordinateSystemMap.get(code).wkt);
},
virtual:{
itemHeight:26
},
height:260
}).data('kendoDropDownList');
```

第四步,创建信息框,使用 disableScrollPropagation 阻止事件派发。

第五步,设置服务地址,新建一个 map,命名为"coordinateSystemMap",用于进行坐标转换。为 anlysis 添加 click 功能,点击后,清空坐标系统地图 coordinateSystemMap,使用 KqGIS.GeometryAnalysis.CoordinateSystemParams 初始化参数 params,令其 id=0,isrecurse 是否递归设置为 true。最后传入 serviceUrl 和 params 参数,调用 geometryAnalysisService 启动查询。

```
    let serviceUrl=service_ip + (isMicroService ? '/geometry-proxy' : '') + '/kqgis/rest/services/geometry';
    let coordinateSystemMap=new Map();
    $('#analysis').on('click',function() {
    coordinateSystemMap.clear();
    let params=new KqGIS.GeometryAnalysis.CoordinateSystemParams({
    id:0,
    isRecurse:true
    });
    L.KqGIS.geometryAnalysisService(serviceUrl).coordinateSystem(params,onSuccess,onFailed);
    });
```

第六步,创建 extractEPSGFromWkt 函数用于从 wkt 中提取 EPSG 信息。使用 split 方法分隔 wkt 中的 AUTHORITY 值,并将最后一位 pop 导出。随后,用空白替换多余的]符号。完成后,用 JSON.parse 将 authority 转为 JavaScript 对象,并使用 join 方法添加。

```
function extractEPSGFromWkt(wkt) {
//获取最后一个 AUTHORITY
let authority=wkt.split('AUTHORITY').pop();
//去掉多余的]符号
authority=authority.replace(/\]/,'');
return JSON.parse(authority).join(':');
```

第七步，创建 getCoordinateSystem，传入 nodes 节点后，对每个节点进行遍历。如果该节点被折叠（存在子节点），则继续递归调用函数访问其子节点。否则，读取节点的 name 和 wkt 并赋值给相应变量。以 name 和 wkt 为参数调用 extractEPSGFromWkt。

```
function getCoordinateSystem(nodes) {
for (let i=0; i < nodes.length; i++) {
let node=nodes[i];
if (node.isFolder) {
getCoordinateSystem(node.childNodes);
} else {
let name=node.name;
let wkt= node.wkt;
coordinateSystemMap.set(extractEPSGFromWkt(wkt),{
id: node.id,
name: name,
wkt: wkt
});
}
}
}
```

如果成功查询，则先调用 getCoordinateSystem 取到 response.result 的子节点信息。然后创建空数组 data，把 coordinateSystemMap.keys 全部导入 data 数组中。设置 kendo.data 的数据源信息，其中 data 属性值为 data 数组，pagesize 页面大小属性值为 40。使用 setDataSource 方法把数据源导入 epsgList 列表中。列表初始值设置为 0。最后调用 setTimeout 方法，把 code、name、wkt 等信息赋值给对应容器，提示"请求成功"，否则提示"请求失败"。

```
let onSuccess=function(response) {
getCoordinateSystem(response.result.result.childNodes);
let data=[];
let keys=[...coordinateSystemMap.keys()];
for (let i=0; i < keys.length; i++) {
data.push({
code: keys[i],
index: i + 1
});
}
let dataSource=new kendo.data.DataSource({
data: data,
```

```
pageSize:40
});
epsgList.setDataSource(dataSource);
epsgList.select(0);
setTimeout(function(){
letcode=epsgList.text();
$('#proj-name').val(coordinateSystemMap.get(code).name);
$('#result').val(coordinateSystemMap.get(code).wkt);
},100)
notification.show({message:'请求成功'},'info');
};
let onFailed=function(){
notification.show({
message:'请求失败'
},'info');
};
});
}
```

运行效果如图7-2所示。

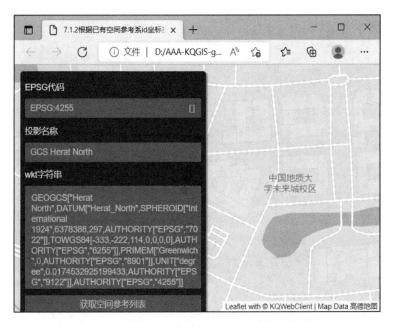

图7-2 根据已有空间参考系ID投影效果预览

7.2 几何拓扑分析

拓扑关系是 GIS 描述要素的空间位置关系,又称位相关系。在实际应用时,有些时候点、线、面各图征间必须保持着某种关系。例如:行政区的范围不能重叠(面的规则)、道路间的线段不能重复(线的拓扑规则)、公车站牌需要在道路上(点的拓扑规则)。因此拓扑是 GIS 中点、线、面图征一些规则与关系的组合,有助于让数据更清楚地仿真真实世界的几何关系,同时降低数字化或编辑上的错误。KQGIS 提供要素拓扑分析,具体步骤如下:

第一步,参考 3.1 创建地图容器并初始化地图,设置地图投影中心,缩放层级,创建 gaodeTilelayer 图层,并调用 addTo 方法将矢量图层(vec)添加到地图中。

第二步,使用 L.Control 类自定义 TopologicalRelatio 控件,使用 DomUtil 类的 create 方法创建拓扑分析后的结果视窗容器,设置"拓扑查询对象 1""拓扑查询对象 2""拓扑关系"容器以及"拓扑关系判定"按钮。随后定义 geometryProjection 函数,以 options 作为输入参数。调用 geometryProjection 方法创建 GeometryProjection 容器,使用 addTo 方法把新建的容器添加到地图的左上角。options 参数设置方法见第五步。

```
L.Control.TopologicalRelation=L.Control.extend({
onAdd: function (map) {
var control_html =    L.DomUtil.create('div','topological_relation');
var template=`<div id="maptools">
<div>
<span>Geometry1</span>
<input id="geometry1" type="text">
</div>
<div>
<span>Geometry2</span>
<input id="geometry2" type="text">
</div>
<div>
<span>拓扑关系</span>
<input id="operator" type="text">
</div>
<button id="button">拓扑关系判定</button>
</div>
`;
setTimeout(function () {
$(".topological_relation").append(template);
```

```
});
return control_html;
},
});
L.control.topologicalRelation=function(opts){
return new L.Control.TopologicalRelation(opts);
};
L.control.topologicalRelation({position:'topleft'}).addTo(map);
```

第三步,初始化待拓扑分析的要素,应包括其属性及样式信息、坐标及投影信息。

```
var geometryA={"type":"FeatureCollection","features":[{"type":"Feature","prop-
erties":{"options":{"stroke":true,"color":"#3388ff","weight":4,"opacity":0.5,"fill":
true,"fillColor":null,"fillOpacity":0.2,"showArea":true,"clickable":true,"_dashArray":
null}},"geometry":{"type":"Polygon","coordinates":[[[114.24896561,30.60113671],
[114.24896561,30.69497956],[114.40108386,30.69497956],[114.40108386,
30.60113671],[114.24896561,30.60113671]]]}}],"proj4":"+proj=longlat+datum=
WGS84+no_defs","prj":"GEOGCS[\"GCS_WGS_1984\",DATUM[\"D_WGS_1984\",
SPHEROID[\"WGS_1984\",6378137.0,298.257223563]],PRIMEM[\"Greenwich\",
0.0],UNIT[\"Degree\",0.0174532925199433],AUTHORITY[\"EPSG\",4326]]"};
```

第四步,使用 L.geoJSON 将要素加入地图,使用 bindPopup 方法绑定相应的 popup 图层,参考 7.1.1 初始化 kendo 下拉列表,设置拓扑分析查询选项及拓扑分析类型选项,使用 disableScrollPropagation 终止事件派发。

```
L.geoJSON(geometryA,{color:'red'}).addTo(map).bindPopup('geometryA');
L.geoJSON(geometryB,{color:'gray'}).addTo(map).bindPopup('geometryB');
L.geoJSON(geometryC,{color:'black'}).addTo(map).bindPopup('geometryC');
L.geoJSON(geometryD,{color:'yellowgreen'}).addTo(map).bindPopup('geometryD');
setTimeout(function(){
//初始化 kendo 容器
$("#geometry1").kendoDropDownList({
dataSource:[
{text:'geometryA',value:JSON.stringify(geometryA)},
{text:'geometryB',value:JSON.stringify(geometryB)},
{text:'geometryC',value:JSON.stringify(geometryC)},
{text:'geometryD',value:JSON.stringify(geometryD)},
],
```

```
dataTextField:"text",
dataValueField:"value",
index:0
}).data("kendoDropDownList");
$("#geometry2").kendoDropDownList({
dataSource:[
{text:'geometryA',value:JSON.stringify(geometryA)},
{text:'geometryB',value:JSON.stringify(geometryB)},
{text:'geometryC',value:JSON.stringify(geometryC)},
{text:'geometryD',value:JSON.stringify(geometryD)},
],
dataTextField:"text",
dataValueField:"value",
index:1
}).data("kendoDropDownList");
$("#operator").kendoDropDownList({
dataSource:[
{text:'相交',value:'intersect'},
{text:'相等',value:'equals'},
{text:'相离',value:'disjoint'},
{text:'接触',value:'touches'},
{text:'相交',value:'crosses'},
{text:'内含',value:'within'},
{text:'包含',value:'contains'},
{text:'重叠',value:'overlaps'},
]
```

第五步，为 button 设置 click 功能，当按钮被按下时，执行函数。首先创建空列表 options。设置 options 的 url 为 rest 服务出图地址。使用 JSON.parse 方法将创建好的 JSON 对象 geometry1 转为 JavaScript 对象。然后使用 getGeomFromFeatureCollection 进行查询，查询结束后，使用 JSON.stringify 将查询结果转为字符串，并命名为"options.geometry1"。同理，创建 options.geometry2。设置 operator 及 checkgeo 参数，进行拓扑分析，读取用户输入的拓扑分析类型及拓扑分析对象，调用拓扑分析服务，若成功，则将返回的 JSON 解码并在结果视窗中显示；否则，提示"请求失败"。GeometryOperator 类的参数如表 7－2 所示。

表 7-2 GeometryOperator 参数介绍

Name	Type	参数定义
operator	String	图形操作类型，必选项
geometry1	GeoJSONObject	options.geometry2 -几何对象 A -几何对象 A,必选项
checkGeo	String	是否检查几何,可选项

```
//拓扑关系判断
$("#button").on("click",function(){
//接口 options
var options={};
options.url=service_ip + "/KQGis/rest/services/jingjin4326";
options.geometry1 = JSON.stringify(getGeomFromFeatureCollection(JSON.parse($("#geometry1").val())));
options.geometry2 = JSON.stringify(getGeomFromFeatureCollection(JSON.parse($("#geometry2").val())));
options.operator=$("#operator").val();
options.checkGeo=true;
var query=newL.kqmap.services.GeometryOperator(options);
query.queryAsync(onsuccess,onfailed);
});
var onsuccess=function(response){
notification.show({
message:"请求成功,判定结果为" + JSON.parse(response).result[0].value
},"info");
};
var onfailed=function(){
notification.show({
message:"请求失败"
},"info");
}
});
}
```

运行结果如图 7-3 所示。

图 7-3 拓扑分析效果预览

7.3 要素缓冲分析

一个面状要素即缓冲要素,点状要素、线状要素和面状要素,被缓冲分析功能处理过后,它们的周围产生一个缓冲区域,该区域即新产生的面状要素。通常,将要素缓冲分析分为单圈缓冲和多圈缓冲。

7.3.1 单圈缓冲

单圈缓冲通常用于分析区位影响半径,其原理是在要素内部或者外部生成一圈指定半径的缓冲图层,要实现单圈缓冲,需进行如下步骤:

第一步,参考 3.1 创建地图容器并初始化地图,设置地图投影中心,缩放层级,创建 gaodeTilelayer 图层,并调用 addTo 方法将矢量图层(vec)添加到地图中。

第二步,添加自定义控件,设置标题,转换结果视窗及转换按钮名称。创建相应函数将他们添加至地图中。

```
L. Control. BufferingAnalysis=L. Control. extend({
onAdd: function (map) {
```

```
var control_html=L.DomUtil.create('div','buffering_analysis');
var template=`<div id="maptools">
<div>
<span>缓冲半径(米)</span>
<input value="100" id="radius" type="text">
</div>
<div>
<button id="analysis">外缓冲分析</button>
<button id="delete">取消外缓冲分析图层</button>
</div>
</div>
`;
setTimeout(function() {
    $(".buffering_analysis").append(template);
});
return control_html;
},
});
L.control.bufferingAnalysis=function(opts) {
    return new L.Control.BufferingAnalysis(opts);
};
L.control.bufferingAnalysis({ position: 'topleft' }).addTo(map);
```

第三步,创建 geo_src 变量,用于初始化数据类型即待缓冲坐标数组。

```
var geo_src={
    type:"Polygon", coordinates:[[[114.561665,30.490096],[114.561665,30.499563],
[114.573338,30.499563],[114.573338,30.490096],[114.561665,30.490096]]]
};
```

第四步,以 geo_src 作为参数传入 L.geoJSON 将要素加入地图,设置其样式及颜色。并使用 fitBounds 适应窗口,参考 7.1.1 初始化 kendo 容器,使用 disableScrollPropagation 阻止事件派发。

```
var geoJSON=L.geoJSON(geo_src,{
style: {
"color": "blue",
"weight": 1
}
```

```
}).addTo(map);
map.fitBounds(geoJSON.getBounds());
setTimeout(function () {
```

第五步,设置 options,其参数如表 7-3 所示。创建 L.kqmap.services.BufferAnalysis 缓冲分析类的对象 query,使用 queryAsync 方法,该方法由子类实例化后调用缓冲分析服务。若成功,将返回 response.result 使用 JSON.parse 解码后,使用 L.geoJSON 类实例化对象 analysisLayer 承接,填写好风格样式后,用 addTo 方法加入地图。同时,设置信息框内容为"请求成功"。

表 7-3 BufferAnalysis 参数介绍

Name	Type	参数定义
url	String	服务器的 url 地址
data	GeoJSONObject	要转换的图形对象,类型为 GeometryCollection 或 Polygon 等 GeoJSON 对象
geoSRS	String	图形的坐标所属的空间参考
outSRS	String	目标空间参考
sideType	String	缓冲区方向(outer:外缓冲区,inner:内缓冲区,left:左缓冲区,right:右缓冲区,both:双缓冲区)
radius	Number	缓冲半径,单位:米,值范围(0,1000000)

```
var analysisLayer=null;
//外缓冲分析
$("#analysis").on("click",function () {
//接口 options
var options={};
options.url=service_ip + "/KQGis/rest/services/jingjin4326";
options.data =JSON.stringify(geo_src);
options.geoSRS='EPSG:4326';
options.outSRS='EPSG:4326';
options.sideType='outer';
options.radius=Number.parseFloat($("#radius").val());
options.version='2.0';
var query=new L.kqmap.services.BufferAnalysis(options);
query.queryAsync(onsuccess,onfailed);
});
var onsuccess=function (response) {
```

```
analysisLayer=L.geoJSON(JSON.parse(response).result,{
style:{
"color":"red",
"weight":1
}
}).addTo(map);
map.flyToBounds(geoJSON.getBounds());
notification.show({
message:"请求成功"
},"info");
```

随后将"缓冲分析"按钮禁用,将"取消缓冲分析图层"按钮启用,并设置"取消缓冲分析图层"按钮的功能为:清空分析图层 analysisLayer,启用"缓冲分析"按钮,禁用删除按钮。若请求失败,则将信息框内容设置为"请求失败",启用"缓冲分析"按钮,禁用"取消缓冲分析图层"按钮。

```
analysisButton.enable(false);
deleteButton.enable(true);
$("#delete").on("click",function(){
map.removeLayer(analysisLayer);
analysisButton.enable(true);
deleteButton.enable(false);
});
};
var onfailed=function(){
notification.show({
message:"请求失败"
},"info");
analysisButton.enable(true);
deleteButton.enable(false);
}
});
}
</script>
```

运行效果如图 7-4 所示。

图 7-4 单圈缓冲效果预览

7.3.2 多圈缓冲

多圈缓冲与单圈缓冲类似，但不同的是多圈缓冲会在要素周围按照一定间距生成多个缓冲区，用于对比不同半径受到影响的大小差异，实现步骤与 7.3.1 基本一致，不同之处在于接口 options 设置中，sidetype 属性应为"both"(深色部分)。

```
$("#analysis").on("click",function () {
//接口 options
var options={};
options.url=service_ip + "/KQGis/rest/services/jingjin4326";
options.data=JSON.stringify(geo_src);
options.geoSRS='EPSG:4326';
options.outSRS='EPSG:4326';
options.sideType ='both';
options.radius=Number.parseFloat($("#radius").val());
options.version='2.0';
var query=new L.kqmap.services.BufferAnalysis(options);
query.queryAsync(onsuccess,onfailed);
});
```

运行效果如图 7-5 所示。

图 7-5　多圈缓冲分析效果预览

7.4　多边形裁剪分析

多边形裁剪为对多边形数据集进行裁剪，包括内部裁剪和外部裁剪。若进行内部裁剪，则被裁剪的多边形数据集在裁剪区范围内的部分被保留到结果数据集中；相反，使用外部裁剪，则保留不在裁剪区范围内的那部分数据到结果数据集中。

以多边形内部裁剪为例：

第一步，参考 7.1.1 进行地图初始化和容器初始化，使用 addTo 方法将创建好的容器及高德地图矢量图层 vec 及矢量注记图层 cva 加入地图中。

第二步，设置 wms 出图地址，使用 L.KqGIS.wmsLayer 实例化 wms 图层对象，设置图层、透明度、样式，使用 addTo 方法加入地图中。使用 L.polygon 类初始化裁剪区域的坐标点，设置其颜色为蓝色，加入地图。

```
var wmsUrl=kqcloud.kqcloud_ip + '/kqcloudserver/wms';
//添加 wms 图层 长株潭
L.KqGIS.wmsLayer(wmsUrl,{
layers: kqcloud.cztPolygon,
opacity: 0.6,
styles: 'M-面-太阳黄'
```

```
}).addTo(map);
L.rectangle(
[
[28.06174966419842,113.09198390965624],
[28.815674224745294,114.46547023778123]
],
{ color: 'blue',fillOpacity: 0 }
).addTo(map);
```

第三步，参考 7.1.1 添加自定义控件，设置标题，转换结果视窗及转换按钮名称。创建相应函数将其添加至地图中。

```
L.Control.OverlayAnalysis=L.Control.extend({
onAdd: function (map) {
var control_html=L.DomUtil.create('div','attr');
var template=`<div id="maptools">
<div>
<span>图层</span>
<input type="text" value="长株潭" readonly>
</div>
<div>
<span>图形(需转换成 wkt 格式)</span>
<input type="text" value="POLYGON((113.09198390965624 28.06174966419842,
114.46547023778123 28.06174966419842,114.46547023778123 28.815674224745294,
113.09198390965624 28.815674224745294,113.09198390965624 28.06174966419842))" readonly>
</div>
<div>
<span>分析结果</span>
<textarea id="result" class="k-textbox"></textarea>
</div>
<button id="query">查询</button>
</div>
`;
control_html.innerHTML=template;
return control_html;
}
});
```

```
L. control. overlayanalysis=function (opts) {
return new L. Control. OverlayAnalysis(opts);
};
L. control. overlayanalysis({ position: 'topleft' }). addTo(map);
```

第四步,创建查询函数 queryDetail,初始化 wkt,layerid 等参数信息。使用 KqGIS. Bigdata. SpatialAnalysis. OverlayDetailParams 类初始化参数列表对象。该类的参数介绍如表 7-4 所示。

表 7-4 OverlayDetailParams 参数介绍

Name	Type	Default	参数定义
url	string		服务全地址
layerId	string		图层 Id
geomList	Array.<string>		可选,图形数组
fields	Object		可选,返回字段
where	string		可选,过滤条件
pageIndex	number	1	可选,分页页码,Default value:1
pageSize	number	10	可选,每页条数,Default value:10
fileType	string		可选,下载文件类型(shp,excel,csv),不为空时下载
returnGeom	boolean	true	可选,返回 geom

```
function queryDetail() {
var wkt =
'POLYGON((113.09198390965624 28.06174966419842,114.46547023778123 28.06174966419842,114.46547023778123 28.815674224745294,113.09198390965624 28.815674224745294,113.09198390965624 28.06174966419842))';
var layerId=kqcloud. cztPolygon. split(':')[1];
const url=kqcloud. kqcloud_ip;
var fields={
"NAME": "NAME",
"the_fid": "the_fid",
geometry: 'clip_geom'
};
var params=new KqGIS. Bigdata. SpatialAnalysis. OverlayDetailParams();
params. geomList=[wkt];
params. layerId=layerId;
params. fields=fields;
```

将 params 作为参数输入调用查询,根据查询状态设置 loading 容器的样式为 display 或 none。若成功返回,则使用 forEach 方法对每个 response 结果的 element 进行遍历。创建 feature 变量存放 element.geometry 的 WKT 值。创建 JSON 变量存放 element.geometry 的 GEoJSON,再将其通过 L.geoJSON 创建对象,设置颜色并添加入地图。在 result 容器中展示 response.content 转换后的 JSON 值。设置 notification 容器的内容为"分析成功",否则提示"分析失败",并设置相应的 loading 状态。

```
L.KqGIS.Bigdata.spatialAnalysisService(url).overlayDetail(
params,
(response) => {
$('#loading').css('display','none');
response = response.result;
if(response.code !== 200){
return;
}
response.content.forEach((element) => {
let feature = new KqGIS.Format.WKT().read(element.geometry);
let JSON = new KqGIS.Format.GeoJSON().write(feature.geometry); // wkt 转成 geoJSON 格式
L.geoJSON(JSON,{ style: { color: 'red' } }).addTo(map); //显示数据
});
```

运行效果如图 7-6 所示。

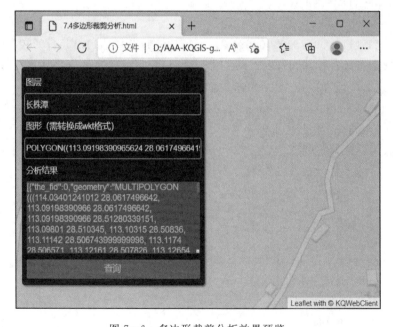

图 7-6　多边形裁剪分析效果预览

7.5 要素叠加

叠加分析是GIS中的一项非常重要的空间分析功能,是指在同一空间参考系统下,通过对两个数据进行的一系列集合运算,产生新数据的过程。这里提到的数据可以是图层对应的数据集,也可以是地物对象。叠加分析的目标是分析在空间位置上有一定关联的空间对象的空间特征和专属属性之间的相互关系。

第一步,参考7.1.1进行地图初始化和容器初始化,使用addTo方法将创建好的容器及高德地图矢量图层加入地图中。参考7.1.1添加自定义控件,设置标题,转换结果视窗及转换按钮名称。创建相应函数将他们添加至地图中。

第二步,使用L.polygon类创建对象srcData,初始化要进行叠加的要素坐标点。

```
let srcData=L.polygon([
[
[30.51836858,114.51777651],
[30.61680016,114.51777651],
[30.61680016,114.69465642],
[30.51836858,114.69465642],
[30.51836858,114.51777651]
```

同样地,初始化另一个要素的坐标点,设置其颜色为红色,加入地图。

```
let destData =L.polygon([
[
[30.46437815,114.61006165],
[30.56381094,114.61006165],
[30.56381094,114.7743073],
[30.46437815,114.7743073],
[30.46437815,114.61006165]
]
],{ color: 'red' }).addTo(map);
```

第三步,调用settimeout方法,执行函数方法,首先,为tolerance容器绑定kendoNimericTextBox富文本框。设置overlayTypeData叠加方法选项。例如,设定文本信息为"交集",以KqGIS.OverlayType类型的INTERSECT属性作为value值。其他选项以同样的方法设置。参考7.1.1完成kendo容器绑定和初始化,使用DisableScrollPropagation方法终止事件派发。

```
setTimeout(function () {
//初始化kendo容器
let tolerance= $('#tolerance').kendoNumericTextBox().data('kendoNumericTextBox');
let overlayTypeData=[
```

```
{ text：'交集',value：KqGIS.OverlayType.INTERSECT },
{ text：'差集',value：KqGIS.OverlayType.ERASE },
{ text：'并集',value：KqGIS.OverlayType.UNION },
{ text：'交集取反',value：KqGIS.OverlayType.SYMMETRICALDIFFERENCE }
```

第四步,设定服务地址,初始化 analysisLayer 为空值。为 analysis 容器添加 click 功能,当该容器被点击时,使用 KqGIS.ServiceSRS 创建 geoSRS 对象,设定其 type 为 EPSG,值为 4326。同样地,创建 destGeoSRS,outSRS 对象,将其值初始化为 EPSG4326。设置 KqGIS.GeometryAnalysis.OverlayParams 的对象 params 用于存放叠加查询的参数。具体参数介绍如表 7-5 所示。

表 7-5 GeometryAnalysis.OverlayParams 参数介绍

Name	Type	Default	参数定义
srcData	KqGIS.GeometryCollection | KqGIS.Geometry.Point | KqGIS.Geometry.MultiLineString | KqGIS.Geometry.MultiPolygon | L.GeoJSON | L.Polygon		源图形对象,类型为 GeometryCollection 或 Point、multilinestring、multipolygon 等 GeoJSON 对象
destData	KqGIS.GeometryCollection | KqGIS.Geometry.Point | KqGIS.Geometry.MultiLineString | KqGIS.Geometry.MultiPolygon | L.GeoJSON | L.Polygon		目标图形对象,类型为 GeometryCollection 或 Point、multilinestring、multipolygon 等 GeoJSON 对象
overlayType	KqGIS.OverlayType		叠置分析类型,交、差、并、交集取反：Intersect、Erase、Union、SymmetricalDifference
tolerance	number	0.00000001	可选容差默认为 0.00000001 米
geoSRS	KqGIS.ServiceSRS		源图形空间参考,支持 EPSG、WKT、proj4 字符串参数,例如 epsg 类型 EPSG：4490
destGeoSRS	KqGIS.ServiceSRS		目标图形空间参考,支持 EPSG、WKT、proj4 字符串参数,例如 epsg 类型 EPSG：4490
outSRS	KqGIS.ServiceSRS		输出图形空间参考,支持 EPSG、WKT、proj4 字符串参数,例如 epsg 类型 EPSG：4490

```
let serviceUrl = service_ip + (isMicroService ? '/geometry-proxy':'') +'/kqgis/rest/services/geometry';
let analysisLayer = null;
//叠置分析
```

```javascript
$('#analysis').on('click',function () {
let geoSRS=new KqGIS.ServiceSRS({
type：KqGIS.ProjectSystemType.EPSG,
value：'4326'
});
let destGeoSRS=new KqGIS.ServiceSRS({
type：KqGIS.ProjectSystemType.EPSG,
value：'4326'
});
let outSRS=new KqGIS.ServiceSRS({
type：KqGIS.ProjectSystemType.EPSG,
value：'4326'
});
let params=new KqGIS.GeometryAnalysis.OverlayParams({
srcData：srcData,
destData：destData,
overlayType：overlayType.value(),
radius：tolerance.value(),
geoSRS：geoSRS,
destGeoSRS：destGeoSRS,
outSRS：outSRS
});
```

填写完成后,使用 L.KqGIS.geometryAnalysisService 方法调用叠加分析。若执行成功,返回 onsuccess 回调函数,使用 result 存放 response.result 的分析结果,设置 result 的类型 type 为 GeometrtColletion。然后使用 geoJSON 类将 result 创建成颜色为绿色、宽度为 1 的要素,用 addTo 方法加入地图。随后将"叠加分析"按钮禁用、"删除叠加区"按钮启用,并设置"删除"按钮的功能为清空分析图层 analysisLayer,启用"叠加分析"按钮,禁用"删除"按钮。若请求失败,则将信息框内容设置为"请求失败",启用"叠加分析"按钮,禁用"删除"按钮。

```javascript
L.KqGIS.geometryAnalysisService(serviceUrl).overlay(params,onSuccess,onFailed);
});
let onSuccess=function (response) {
let result=response.result.result;
result.type='GeometryCollection';
analysisLayer=L.geoJSON(result,{ style：{ 'color': 'green','weight': 1 } }).addTo(map);
notification.show({ message：'请求成功' },'info');
```

```
analysisButton.enable(false);
deleteButton.enable(true);
$('#delete').on('click',function () {
map.removeLayer(analysisLayer);
analysisButton.enable(true);
deleteButton.enable(false);
});
};
let onFailed=function () {
notification.show({
message：'请求失败'
},'info');
analysisButton.enable(true);
deleteButton.enable(false);
};
});
}
</script>
```

运行效果如图 7-7 所示。

图 7-7　要素叠加效果预览

7.6 计算要素实地周长

计算要素实地周长是 GIS 系统中最基本的应用之一，其原理是根据地图上丈量的尺寸再乘以比例尺的倍数，计算实地的周长，具体实现步骤如下。

第一步，参考 7.1.1 进行地图初始化和容器初始化，将创建好的容器及高德地图矢量图层使用 addTo 方法加入地图中。添加自定义控件，设置标题，转换结果视窗及转换按钮名称。创建相应函数将它们添加至地图中。

```
L.Control.LengthCalculation=L.Control.extend({
onAdd: function (map) {
var control_html =   L.DomUtil.create('div','length_calculation');
var template=`<div id="maptools">
<input id="length" type="text" class="k-input k-textbox">
<button id="calculate">计算长度</button>
</div>`;
setTimeout(function () {
 $(".length_calculation").append(template);});
return control_html;},});
L.control.lengthCalculation = function(opts) {return new L.Control.LengthCalculation(opts);};
L.control.lengthCalculation({ position: 'topleft' }).addTo(map);
```

第二步，初始化坐标点数组 coordinates，将其作为参数传入 L.polyline 类的构造函数生成折线对象 polyline，设置其颜色为红色，使用 addTo 方法加入地图并使用 fitBounds 调整地图显示范围，其中范围大小由 polyline.getBounds() 决定。

```
var coordinates=[[30.490096,114.561665],[30.480629,114.573338],[30.499563,114.585011]];
var polyline=L.polyline(coordinates,{color: 'red'}).addTo(map);
map.fitBounds(polyline.getBounds());
```

第三步，初始化 kendobutton 及 notification，设置容器的样式及调用的函数，使用 disableScrollPropagation 终止事件派发。

第四步，为 calculate 绑定 click 功能，当该容器被点击时，初始化 options 列表为空值、options.url 为服务地址。将 options 作为参数，调用长度计算服务 LengthCalculate，其参数如表 7-6 所示。

表 7-6 LengthCalculate 参数介绍

Name	Type	参数定义
url	String	服务器的 url 地址
data	GeoJSONObject	要计算面积或长度的图形对象，必选项。 类型为 GeometryCollection 或 Polygon 等 GeoJSON 对象
geoSRS	String	图形的坐标所属的空间参考，必选项
outSRS	String	计算面积所使用的空间参考，必选项

```
//计算长度
$("#calculate").on("click",function () {
varoptions={};options.url=service_ip + "/KQGis/rest/services/jingjin4326";options.data='{"type":"Polyline","coordinates":[[[114.561665,30.490096],[114.573338,30.480629],[114.585011,30.499563]]]}';options.geoSRS='EPSG:4326';options.outSRS='EPSG:4326';var query=new L.kqmap.services.LengthCalculate(options);
```

填写完 options 之后使用 queryAsync 异步方法调用查询。若执行成功，返回 onsuccess 回调函数，使用 JSON.parse 将 response 对象转为 JSON 并取其 result 结果的第一位的值，在后面加上"米"后，传入 length 容器中。然后使用 flyToBounds 调整地图显示范围，其中范围大小由 polyline.getBounds() 决定。并设置 notification 信息框的 message 为"请求成功"，否则提示"请求失败"。

```
query.queryAsync(onsuccess,onfailed);});
var onsuccess=function (response) { $("#length").val(JSON.parse(response).result[0].value + " 米");map.flyToBounds(polyline.getBounds());
```

运行结果如图 7-8 所示。

图 7-8 计算要素实地周长

7.7 计算要素实地面积

计算要素实地面积是 GIS 系统中最基本的应用之一,其原理是根据地图上丈量的尺寸再乘以比例尺的倍数的平方,计算要素实地面积的步骤与 7.6 基本一致,但在 options 设置中需要将 data 类型改为 polygon,且查询时调用 AreaCalculate 接口,用到的参数如表 7-7 所示。

表 7-7 AreaCalculate 参数介绍

Name	Type	参数定义
url	String	服务器的 url 地址
data	GeoJSONObject	要计算面积或长度的图形对象,必选项。 类型为 GeometryCollection 或 Polygon 等 GeoJSON 对象
geoSRS	String	图形的坐标所属的空间参考,必选项
outSRS	String	计算面积所使用的空间参考,必选项

运行结果如图 7-9 所示。

图 7-9 计算要素实地面积效果预览

8 KQGIS 空间分析

8.1 图层坐标投影转换

图层投影转换用于解决 GIS 系统中图层坐标数据的统一问题,例如以 WGS84 坐标系为标准的图层及以西安 80 坐标系为标准图层互相转换。KQGIS 提供根据空间参考系信息投影和根据已有空间参考系 ID 投影的转换方法。

8.1.1 根据空间参考系信息投影

根据空间参考系信息进行图层投影转换的步骤如下。

第一步,进行地图初始化和容器初始化,使用 addTo 方法将创建好的容器及高德地图矢量图层及矢量注记图层加入地图中。添加自定义控件,设置标题,转换结果视窗及转换按钮名称。创建相应函数将它们添加至地图中。

```
L.Control.QueryFieldInfo=L.Control.extend({
onAdd: function (map) {
var control_html=L.DomUtil.create('div','query_dataset_info');
var template=<div id="maptools">
<div>
<span>数据源</span>
<input type="text" value="KQSpatialDB202206241700" readonly>
</div>
<div>
<span>原始数据集(坐标系 EPSG:4326)</span>
<input type="text" value="CAPTITAL_P" readonly>
</div>
<div>
<span>EPSG 列表</span>
<input id='epsg-code' type='text'>
</div>
<div>
```

```
<span>转换后数据集</span>
<input  type="text" value="CAPTITAL_P_TRANSFER" readonly>
</div>
<button id="query">坐标转换</button>
</div>
;
setTimeout(function () {
 $('.query_dataset_info').append(template);
});
return control_html;
}});
L.control.queryFieldInfo=function (opts) {
return new L.Control.QueryFieldInfo(opts);
};
L.control.queryFieldInfo({position: 'topleft'}).addTo(map);
```

第二步，设置出图地址 jingjinUrl，将其作为参数调用 L.KqGIS.tileMapLayer 创建图层，设置图层 id 及格式，用 addTo 方法加入地图。设置数据地址 dataUrl。

```
let jingjinUrl =
service_ip + (isMicroService ? '/jingjin3857-proxy' : '') + '/kqgis/rest/services/jingjin3857/map';
//添加图层
L.KqGIS.tileMapLayer(jingjinUrl,{
layerIds: [1,2,3,4,5,6,7],
format: 'image/png'
}).addTo(map);
let dataUrl =
service_ip + (isMicroService ? '/jingjin_data-proxy' : '') + '/kqgis/rest/services/jingjin_data/data';
```

第三步，使用 setTimeout 方法，自动调用函数，首先为 query 容器绑定"kendoButton"按钮，当按钮被按下时，创建 KqGIS.Data.DataInfoParams 数据信息参数对象 inputData，同时初始化其 datasourceName 和 datasetName 的值。同样地，创建 outputData，初始化有关参数。

```
setTimeout(function () {
//初始化 kendo 容器
$('#query')
.kendoButton({
```

```
click: function () {
let inputData = new KqGIS.Data.DataInfoParams({
datasourceName: 'KQSpatialDB202206241700',
datasetName: 'CAPITAL_P'
});
let outputData = new KqGIS.Data.DataInfoParams({
datasourceName: 'KQSpatialDB202206241700',
datasetName: 'CAPITAL_P_TRANSFER'
});
```

使用 KqGIS.Data.CoordTransferParams 创建参数列表对象 params，将 inputdata 和 outputdata 作为参数输入，该类所有的参数如表 8-1 所示。按照需求填写完相关参数后，传入 dataurl 及 params 调用 coordTransfer 进行查询，设置查询成功和失败的回调函数。

表 8-1 CoordTransferParams 参数介绍

Name	Type	Default	参数定义
inputData	KqGIS.Data.DatasetInfoParams		待转换的数据
outputData	KqGIS.Data.DatasetInfoParams		可选，转换结果所属表，如果为空，则输出到全局数据源下面
appendMode	boolean	true	可选，数据添加的模式，追加(true)或覆盖(false)，默认追加；对 output 所指向的内容存在的情况，如果不存在，则不予关心
outSRS	KqGIS.ServiceSRS		设置坐标投影转换的目标空间参考

```
letparams = new KqGIS.Data.CoordTransferParams({
inputData: inputData,
outputData: outputData,
appendMode: false,
outSRS: new KqGIS.ServiceSRS({
type: KqGIS.ProjectSystemType.EPSG,
value: '3857'
})})
L.KqGIS.datasetService(dataUrl).coordTransfer(params, onsuccess, onfailed);
}})
.data('kendoButton');
```

为 epsg-code 容器绑定"kendoDropDownList"下拉列表,设置数据文本域为 code、值域为 index、查询条件 filter 为 contains,为 notification 容器绑定 kendoNotification 用于展示查询消息。设置完其样式属性后使用 disableScrollPropagation 终止事件派发。

```
let epsgList = $('#epsg-code')
.kendoDropDownList({
dataTextField: 'code',
dataValueField: 'index',
filter: 'contains',
virtual: {
itemHeight: 26
},
height: 260
})
.data('kendoDropDownList');
var notification = $('#notification')
.kendoNotification({
position: {pinned: true, bottom: 12, right: 12},
autoHideAfter: 2000,
stacking: 'up',
templates: [
{
type: 'info',
template: '<div>#= message #</div>'
}
]
})
.data('kendoNotification');
L.DomEvent.disableScrollPropagation($('#maptools')[0]).disableClickPropagation($('#maptools')[0]);
```

第四步,设置上一步中查询成功的回调函数,该函数会创建一个 KqGIS.Data.GetFeatureListParams 类的查询参数对象,其参数如表 8-2 所示。

表 8-2 GetFeatureListParams 参数介绍

Name	Type	Default	参数定义
datasourceName	string		可选,数据源名称
datasourceID	string		可选,数据源 ID。与 datasourceName 参数功能相同,用于确定数据源

续表 8-2

Name	Type	Default	参数定义
connInfo	KqGIS.Data.ConnectionInfoParams		可选,数据源连接信息,动态打开数据源。功能与其他 datasourceID 和 datasourceName 参数类似,优先级高于其他。该参数用于全局通用数据服务,根据临时连接信息动态打开数据源
datasetName	string		可选,数据集名称。需要与 datasourceName 或 datasourceID 参数组合使用,确定唯一数据集
datasetID	string		可选,数据集 ID。可替代 datasourceName、datasourceID、datasetName 的组合,独自确定唯一数据集。使用 options.connInfo 参数时,本参数无效,请使用 options.datasetName
layerName	string		可选,图层名称,地图服务下生效。可替代 datasourceName、datasourceID、datasetName、datasetID 参数,独自确定唯一数据集
layerId	string		可选,图层 ID,地图服务下生效。与 layerName 参数功能相同,二选一,优先级高于 layerName
fromIndex	number	0	可选,查询结果的最小索引号,整数。默认值是 0,如果该值大于查询结果的最大索引号,则查询结果为空
toIndex	number		可选,查询结果的最大索引号,整数。如果该值大于查询结果的最大索引号,则以查询结果的最大索引号为终止索引号。如果设置了此值,则它必须大于 fromIndex,如果小于 toIndex,则查询失败;设置了无效的值,则下面默认为 19,没设置此值时,返回所有,可能会比较慢
returnContent	boolean	false	可选,是否返回要素信息(属性和图形信息),默认为 false,只返回要素的 id。true 时,返回要素信息
hasGeometry	boolean	false	可选,是否返回要素图形信息,默认为 false,不返回要素图形信息,true 时,返回要素图形信息。在 returnContent 为 true 时,该参数有效
outSRS	KqGIS.ServiceSRS		可选,结果数据集的空间参考

```
var onsuccess=function (response) {
let params=new KqGIS.Data.GetFeatureListParams({
datasourceName: 'KQSpatialDB202206241700',
datasetName: 'CAPITAL_P_TRANSFER',
returnContent: true,
```

```
hasGeometry: true,
outSRS: new KqGIS.ServiceSRS({
type: KqGIS.ProjectSystemType.EPSG,
value: '4326'
})
});
```

设置完参数后,参考上一步,传入 dataUrl 及 params 调用 getFeatureList。若成功,则用 L.geoJSON 将 response.result.result.features 转为要素对象并添加入地图,并在 notification 中提示"请求成功"。否则执行失败,回调函数,提示请求"失败"。

```
L.KqGIS.dataFeatureService(dataUrl).getFeatureList(
params,
(response) => {
L.geoJSON(response.result.result.features).addTo(map);
},
onfailed
);
notification.show({ message:'请求成功' },'info');
};
var onfailed=function () {
notification.show({ message:'请求失败' },'info');
};
```

第五步,设置服务地址 analysisiUrl,创建 coordinateSystemMap,使用 setTimeout 方法自动调用函数,创建 KqGIS.GeometryAnalysis.CoordinateSystemParams 的 params 列表,将 params 及 analysisUrl 作为参数传入并调用 coordinateSystem 进行查询。

```
let analysisUrl= service_ip + (isMicroService ? '/geometry-proxy' : '') + '/kqgis/rest/services/geometry';
let coordinateSystemMap=new Map();
setTimeout(function () {
let params=new KqGIS.GeometryAnalysis.CoordinateSystemParams({
id: 0,
isRecurse: true
});
L.KqGIS.geometryAnalysisService(analysisiUrl).coordinateSystem(params,onCoorSuccess,onCoorFailed);
});
```

创建 extractEPSGFromWkt 函数用于从 wkt 中提取 EPSG 信息。使用 split 方法分隔 wkt 中的 AUTHORITY 值,并将最后一位 pop 导出。随后,用空白替换多余的]符号。完成后,用 JSON.parse 将 authority 转为 JavaScript 对象,并使用 join 方法添加。

```javascript
function extractEPSGFromWkt(wkt) {
//获取最后一个 AUTHORITY
let authority = wkt.split('AUTHORITY').pop();
//去掉多余的]符号
authority = authority.replace(/\]/,'');
return JSON.parse(authority).join(':');
}
```

创建 getCoordinateSystem,传入 nodes 节点后,对每个节点进行遍历。如果该节点被折叠(存在子节点),则继续递归调用函数访问其子节点。否则,读取节点的 name 和 wkt 并赋值给相应变量。以 name 和 wkt 为参数调用 extractEPSGFromWkt。

```javascript
function getCoordinateSystem(nodes) {
for (let i=0; i < nodes.length; i++) {
let node = nodes[i];
if (node.isFolder) {
getCoordinateSystem(node.childNodes);
} else {
let name = node.name;
let wkt = node.wkt;
coordinateSystemMap.set(extractEPSGFromWkt(wkt),{
id: node.id,
name: name,
wkt: wkt
});
}}}
```

如果成功查询,则先调用 getCoordinateSystem 取到 response.result 的子节点信息。然后创建空数组 data,把 coordinateSystemMap.keys 全部导入 data 数组中。设置 kendo.data 的数据源信息,其中 data 属性值为 data 数组,初始 page 为 1,pagesize 页面大小属性值为 20000。使用 setDataSource 方法把数据源 dataSource 导入 epsgList 列表中,并使用 select 方法把列表初始值 dataItem 设置为 EPSG:3857。最后提示"查询成功",否则提示"查询失败"。运行效果如图 8-1 所示。

图 8-1　根据空间参考系信息投影

8.1.2　根据已有空间参考系 ID 投影

根据已有空间参考系进行图层投影转换的步骤和 8.1.1 中介绍的过程基本相同，但在容器初始化的过程中需将 outSRS 参数的 type 改为 KqGIS.ProjectSystemType.WKT。

```
let params=new KqGIS.Data.CoordTransferParams({
inputData: inputData,
outputData: outputData,
appendMode: false,
outSRS: newKqGIS.ServiceSRS({
type: KqGIS.ProjectSystemType.WKT,
value: coordinateSystemMap.get(epsgList.text()).wkt
})
});
```

运行效果如图 8-2 所示。

图 8-2 根据已有参考系 id 投影

8.2 图层拓扑查错

数据生产中,其中一个核心问题便是数据完整性,KqGIS 中对数据完整性的检查提供了拓扑查错接口,它可以验证地理数据是否正确完整。

第一步,进行地图初始化和容器初始化,将创建好的容器及高德地图矢量图层及矢量注记图层使用 addTo 方法加入地图中。

第二步,使用 L.geoJSON 创建要素,设置其类型 type 为多边形,设置 coordinates 数组存储其坐标范围,使用 addTo 方法将其加入地图。

```
let geoJSON=L.geoJSON({
'type': 'Polygon',
'coordinates': [
[
[100.0,30.0],
[102.0,30.0],
[102.0,32.0],
[100.0,32.0],
[102.0,28.0]
]
]}).addTo(map);
```

第三步，使用 setTimeout，为 notification 容器绑定 kendoNotification 消息展示框，设置完其样式属性后使用 disableScrollPropagation 终止事件派发。

第四步，设置服务地址 serviceUrl，为 analysis 容器绑定点击事件，当事件触发时，使用 KqGIS.ServiceSRS 类创建对象 geoSRS，设置其类型为 EPSG、值为 4326。随后调用 KqGIS.GeometryAnalysis.GeometryCheckParams 类实例化查询列表 params，该类参数如表 8-3 所示。

表 8-3 GeometryCheckParams 类参数介绍

Name	Type	参数定义
data	KqGIS.GeometryCollection | KqGIS.Geometry.MultiPolygon | L.GeoJSON | L.Polygon	需要分析的图形对象，类型为 GeometryCollection 或 multi-polygon 等 GeoJSON 对象
geoSRS	KqGIS.ServiceSRS	图形的坐标所属的空间参考，支持 EPSG、WKT、proj4 字符串参数，例如 epsg 类型 EPSG:4490
params	KqGIS.ToleranceParams	可选，图形检查参数 n，PointLimit：容差，AreaLimit：最小面积

```
let serviceUrl=service_ip + (isMicroService ? '/geometry-proxy':'') + '/kqgis/rest/services/geometry';
//几何属性服务
$('#analysis').on('click',function () {
let geoSRS=new KqGIS.ServiceSRS({
type: KqGIS.ProjectSystemType.EPSG,
value: '4326'
});
let params=new KqGIS.GeometryAnalysis.GeometryCheckParams({
data: geoJSON,
geoSRS: geoSRS,
params: {
'PointLimit': 0.0000001,
'AreaLimit': 0.01
}
```

将 serviceUrl、params 作为参数传入，调用 L.KqGIS.geometryAnalysisService 进行查询，设置回调函数。

```
L.KqGIS.geometryAnalysisService(serviceUrl).geometryCheck(params,onSuccess,onFailed);
});
```

若查询成功,则创建变量 result 保存 response.result.result 的值,并使用 JSON.stringify 将 result 对象转化为字符串,再使用 formatJSON 格式化解析。解析完成后,将其值作为 result 的 val 属性进行展示,然后设置信息框中的内容为"请求成功"。否则设置为"请求失败"。

```
let onSuccess = function (response) {
let result = response.result.result;
$('#result').val(formatJSON(JSON.stringify(result)));
notification.show({ message: '请求成功' }, 'info');
};
let onFailed = function () {
notification.show({
message: '请求失败'
}, 'info');
};
```

运行效果如图 8-3 所示。

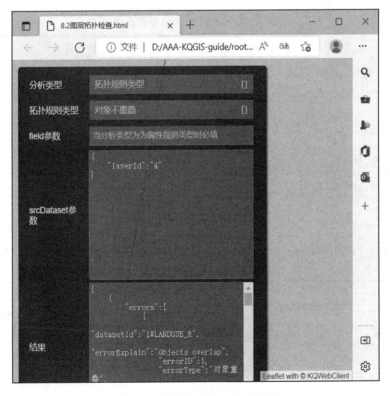

图 8-3 图层拓扑检查效果预览

8.3 图层缓冲分析

图层缓冲分析用于两个或多个不同的图层之间进行缓冲区分析,包括左缓冲、右缓冲、双缓冲等方式,具体实现方法如下。

第一步,设置出图地址和服务地址。完成地图初始化及容器初始化,并将它们用 addTo 方法加入地图中。

第二步,使用 setTimeout 方法,为 radius 容器绑定 kendoNumericTextBox 富文本编辑框,设置 sideTypeData 数组,存放待选择的选项。为 sideType 容器绑定"kendoDropDownList"下拉列表,设置其数据源为 sideTypeData。同样地,设置 headType、cornerType 容器的相关参数并绑定下拉列表,使用 disableScrollPropagation 终止事件派发。

第三步,创建 analysisLayer 变量,设置其值为空值。为 analysis 容器绑定鼠标点击事件,当事件触发时,将 loading 容器的样式设置为 display/block。使用 KqGIS.ServiceSRS 类创建对象 geoSRS,设置其类型为 EPSG、值为 4326。用 KqGIS.SpatialAnalysis.BufferParams 类实例化查询列表 params。该类参数如表 8-4 所示。

表 8-4 BufferParams 参数介绍

Name	Type	Default	参数定义
dataset	KqGIS.SpatialAnalysis.DataSetInfoMapParams | KqGIS.SpatialAnalysis.DataSetInfoMapParams		源数据,地图服务配置的用 DataSetInfoMapParams,数据服务配置的用 DataSetInfoDataParams
headType	KqGIS.HeadType	Round	可选,缓冲末端类型。Flat 平头,Square 方头,Round 圆头
cornerType	KqGIS.CornerType	Round	可选,缓冲凸角类型。Sharp 尖角,Flat 平角,Round 圆角
dissolve	boolean	false	可选,是否融合,默认为 false
sideType	KqGIS.BufferSideType		缓冲类型。AroundPoint 点的周围 Left 前进方向的左边(说明:对于面图形,将往面外缓冲)Right<前进方向的右边(说明:对于面图形,将往面内缓冲)Both 线或者面的两边
radius	number		缓冲半径(单位:米)
outPutType	number	3	可选,分析结果输出类型,1 代表输出到全局数据源;2 代表输出到当前数据服务指定数据源的数据集;3 代表输出"GeometryCollection"或"FeatureCollection"的 GeoJson 对象
filter	KqGIS.FilterParams		可选,数据过滤条件
outSRS	KqGIS.ServiceSRS		可选,生成数据集的空间参考
aysn	boolean	false	可选,是否异步执行,如果异步执行,直接返回任务 id,通过任务状态查询返回结果。默认同步。当 outPutType 指定为 3 时,该参数无效

续表 8 - 4

Name	Type	Default	参数定义
outDataset	KqGIS. SpatialAnalysis. DataSetInfoDataParams | KqGIS. SpatialAnalysis. DataSetInfoMapParams		可选,结果保存路径。当 outPutType 指定为 2 时,该参数有效

```
let analysisLayer = null;
//缓冲分析
$('#analysis').on('click', function () {
$("#loading").css("display","block");
let geoSRS = new KqGIS.ServiceSRS({
type: KqGIS.ProjectSystemType.EPSG,
value: '4326'
});
let params = new KqGIS.SpatialAnalysis.BufferParams({
dataset: { "layerid": "1" },
sideType: sideType.value(),
radius: radius.value(),
outSRS: geoSRS,
headType: headType.value(),
cornerType: cornerType.value(),
outPutType: 3,
filter: new KqGIS.SQLFilterParams({ where: "NAME='尊贵高速'" }),
dissolve: true
});
```

第四步,将 serviceUrl、params 作为参数传入,调用 L. KqGIS. spatialAnalysisService 进行缓冲分析,设置回调函数。

```
L.KqGIS.spatialAnalysisService(serviceUrl).buffer(params, onSuccess, onFailed);
});
```

当查询成功时,设置 loading 容器的样式设置为 display/block。创建变量 result 保存 response.result.result 的值,将其作为参数使用 L. geoJSON 创建 geoJSON 要素,设置要素的颜色、宽度、透明度等属性后,用 analysisLayer 变量承接,使用 addTo 方法加入地图。然后设置信息框中的内容为"请求成功"。将 analysisButton 禁用、deleteButton 启用。当"删除"按钮被按下时,使用 map. removeLayer 方法移除 analysisLayer 缓冲图层,将 analysisButton 启用、deleteButton 禁用。否则,设置信息框中的内容为"请求失败",将 analysisButton 启用、

deleteButton 禁用。

```javascript
    let onSuccess=function(response){
     $("#loading").css("display","none");
    let result=response.result.result;
    analysisLayer=L.geoJSON(result,{style:{color:'red',weight:1,fillOpacity:0.5}
}).addTo(map);
    notification.show({message:'请求成功'},'info');
    analysisButton.enable(false);
    deleteButton.enable(true);
     $('#delete').on('click',function(){
    map.removeLayer(analysisLayer);analysisButton.enable(true);deleteButton.enable(false);
    });
    };
    let onFailed=function(){
     $("#loading").css("display","none");notification.show({message:'请求失败'
},'info');analysisButton.enable(true);deleteButton.enable(false);
    };
    });
    </script>
```

运行效果如图 8-4 所示。

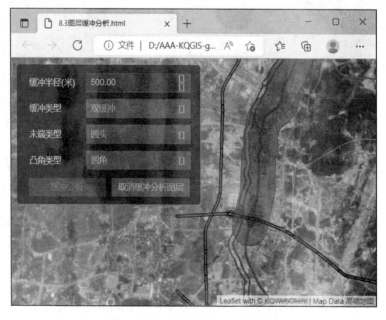

图 8-4 图层缓冲分析效果预览

8.4 图层叠置分析

叠置分析是 GIS 用户经常用以提取数据的手段之一。地图的叠置,按直观概念就是将两幅或多幅地图重叠在一起,产生新数据层和新数据层上的属性,新数据层或新空间位置上的属性就是各叠置地图上相应位置处各属性的函数。下面将介绍图层叠置分析,包括图形与图层叠置分析、图层与图层叠置分析。

8.4.1 图形与图层叠置分析

图形与图层叠置分析的过程如下。

第一步,分别设置图层的出图地址和服务地址。完成地图初始化及容器初始化,并用 addTo 方法将它们加入地图中。

第二步,使用 L.polygon 创建多边形要素,设置宽度和颜色属性后,使用 addTo 方法加入地图。随后调用 setTimeout 方法自动调用函数,当该函数执行时,设置 overlayTypeData 数组,存放叠置分析选项,包括名称和值。为 overlay-type 绑定"kendoDropDownList"下拉列表,设置其数据源为 overlayTypeData。

```
L.polygon([[
[39.5,117.3],
[38.5,117.3],
[38.5,116.3],
[39.5,116.3],
[39.5,117.3]
]],{
weight:4,
color:'blue',
}).addTo(map);
setTimeout(function(){
//初始化 kendo 容器
let overlayTypeData=[
{ text:'交集',value:KqGIS.OverlayType.INTERSECT },
{ text:'差集',value:KqGIS.OverlayType.ERASE },
{ text:'并集',value:KqGIS.OverlayType.UNION },
{ text:'交集取反',value:KqGIS.OverlayType.SYMMETRICALDIFFERENCE }
];
let overlayType=$('#overlay-type').kendoDropDownList({
dataTextField:'text',
```

```
dataValueField: 'value',
dataSource: overlayTypeData,
index: 0
}).data('kendoDropDownList');
```

为 analysis、delete 容器绑定 kendoButton，设置 delete 容器的初始状态为 false 禁用。为 notification 容器绑定 kendoNotification 消息展示框，设置其相关样式信息使用 disableScroll-Propagation 终止事件派发。

第三步，创建 analysisLayer 变量，设置其值为空值。为 analysis 容器绑定鼠标点击事件，当事件触发时，将 loading 容器的样式设置为 display/block。创建 operDataset 变量，使用 L.geoJSON 创建要素，设置要素类型为多边形，设置要素边界的坐标点数组。将创建好的要素作为 operDataset 的 geometry 属性。使用 KqGIS.ServiceSRS 类创建对象 geoSRS，设置其类型为 EPSG、值为 4326。用 KqGIS.SpatialAnalysis.OverlayParams 类实例化查询列表 params。该类参数如表 8-5 所示。

表 8-5 OverlayParams 参数介绍

Name	Type	Default	参数定义
srcDataset	KqGIS.SpatialAnalysis.DataSetInfoMapParams | KqGIS.SpatialAnalysis.DataSetInfoDataParams		源数据，地图服务配置的用 DataSetInfoMapParams，数据服务配置的用 DataSetInfoDataParams
srcDatasetFilter	KqGIS.FilterParams		可选，源数据过滤条件
operDataset	KqGIS.SpatialAnalysis.DataSetInfoMapParams | KqGIS.SpatialAnalysis.DataSetInfoDataParams | KqGIS.SpatialAnalysis.DataSetInfoGeometryParams		操作数据 地图服务配置 DataSetInfoMapParamsDataSetInfoGeometryParams，数据服务配置的用 DataSetInfoDataParams
operDatasetFilter	KqGIS.FilterParams		可选，操作数据过滤条件
operateType	KqGIS.BooleanOperationType		操作类型，枚举值
minArea	number		可选，面积容差（单位：平方米），面积小于 minArea 的 geometry 将被过滤掉
minLength	number		可选，长度容差（单位：米），长度小于 minLength 的 geometry 将被过滤掉
outSRS	KqGIS.ServiceSRS		可选，生成数据集的空间参考

续表 8-5

Name	Type	Default	参数定义
resultFieldName	string		可选,重算结果图形(可能是面也可能是线)面积或长度赋值到 resultFieldName 指定的字段中去,如果 resultFieldName 字段不存在,则新建字段
ellipsoidal	boolean	false	可选,是否为椭球面积,默认为 false;为 true 时返回为目标空间参考下的椭球面积,为 false 时返回其几何面积(图形本身的面积)
sourceDatasetFields	Array.<string>		可选,源数据集中要复制的字段。配合 operateDatasetFields 使用,与 outFields 互斥
operateDatasetFields	Array.<string>		可选,操作数据集中要复制的字段。配合 sourceDatasetFields 使用,与 outFields 互斥
outFields	KqGIS.AttributeInheritanceType		可选,叠置结果字段,结果要素属性继承方式.
outPutType	number	3	可选,分析结果输出类型,1 代表输出到全局数据源;2 代表输出到当前数据服务指定数据源的数据集;3 代表输出 "GeometryCollection"或"FeatureCollection"的 GeoJSON 对象
outDataset	KqGIS.SpatialAnalysis.DataSetInfoDataParams		可选,叠置结果保存路径。当 outPutType 指定为 2 时,该参数有效
returnCount	number		可选,要求返回结果的最大记录数。当 outPutType 指定为 3 时,该参数有效
aysn	boolean	false	可选,是否异步执行,如果异步执行,直接返回任务 id,通过任务状态查询返回结果。默认同步,值为 false。当 outPutType 指定为 3 时,该参数无效

```
let analysisLayer=null;
//叠置分析
$('#analysis').on('click',function () {
$("#loading").css("display","block");
let operDataset={
"geometry": L.geoJSON({
"type": "Polygon",
"coordinates": [[
```

```
        [117.3,39.5],
        [117.3,38.5],
        [116.3,38.5],
        [116.3,39.5],
        [117.3,39.5]]]}),
    "geoSRS": new KqGIS.ServiceSRS({
    type: KqGIS.ProjectSystemType.EPSG,value:'4326'
    })}
    let params=new KqGIS.SpatialAnalysis.OverlayParams({srcDataset: { "layerid": "4"
},operDataset: operDataset,operateType: overlayType.value(),outPutType: 3,outFields:
KqGIS.AttributeInheritanceType.FromSource
    });
```

第四步,将 serviceUrl、params 作为参数传入,调用 L. KqGIS. spatialAnalysisService 进行叠置分析,设置回调函数。

```
    L.KqGIS.spatialAnalysisService(serviceUrl).overlay(params,onSuccess,onFailed);
```

当查询成功时,将 loading 容器的样式设置为 display/block。创建变量 result 保存 response.result.result 的值,设置 result 的 type 为 MultiPolygon,将其作为参数使用 L.geoJSON 创建 geoJSON 要素,设置要素的颜色、宽度、透明度等属性后,用 analysisLayer 变量承接,使用 addTo 方法加入地图。然后设置信息框中的内容为"请求成功"。将 analysisButton 禁用、deleteButton 启用。当"删除"按钮被按下时,使用 map.removeLayer 方法移除 analysisLayer 缓冲图层,将 analysisButton 启用、deleteButton 禁用。

否则设置信息框中的内容为"请求失败",将 analysisButton 启用、deleteButton 禁用。

```
    let onSuccess=function (response) {
    $("#loading").css("display","none");
    let result=response.result.result;
    result.type='MultiPolygon';
    analysisLayer= L.geoJSON(result,{ style: { 'color': 'red','weight': 3 } }).addTo
(map);
    notification.show({ message:'请求成功' },'info');
    analysisButton.enable(false);
    deleteButton.enable(true);
    $('#delete').on('click',function () {
    map.removeLayer(analysisLayer);
    analysisButton.enable(true);
    deleteButton.enable(false);
```

```
});
};
let onFailed=function(){
$("#loading").css("display","none");
notification.show({
message：'请求失败'
},'info');
analysisButton.enable(true);
deleteButton.enable(false);
};
});
}
</script>
```

运行效果如图 8-5 所示。

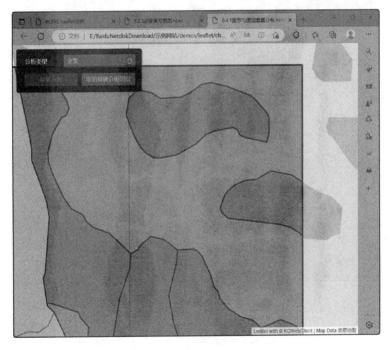

图 8-5　图形与图层叠置分析运行效果

8.4.2　图层与图层叠置分析

图层与图层叠置分析的步骤与上一节类似，但在实例化 params 时，无须设置要素坐标范围，但需要使用 KqGIS.SpatialAnalysis.OverlayParams 初始化设置 srcDataset 和 operDataset 两个不同的图层数据。

```
let params=new KqGIS.SpatialAnalysis.OverlayParams({
srcDataset：{"layerid"："3"},
operDataset：{"layerid"："4"},
operateType：overlayType.value(),
outPutType：3,
outFields：KqGIS.AttributeInheritanceType.FromSource
});
```

运行效果如图 8-6 所示。

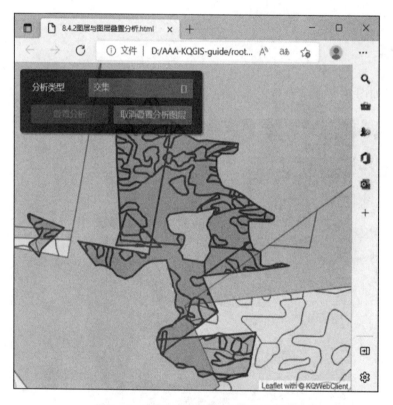

图 8-6　图层与图层叠置分析效果预览

8.5　路径分析

　　路径分析是 GIS 中最基本的功能,其核心是对最佳路径和最短路径的求解,从网络模型的角度看,最佳路径求解就是在指定网络的两结点间找一条阻碍强度最小的路径。最佳路径的产生基于网线和结点转角(如果模型中结点具有转角数据的话)的阻碍强度。

具体实现方式如下。

第一步,完成地图初始化及容器初始化,并用 addTo 方法将它们加入地图中。

第二步,设置道路图层的出图地址 roadUrl 和服务地址 serviceUrl,将其作为参数使用 KqGIS.tileMapLayer 初始化图层,使用 addTo 方法将图层加入地图。

```
//加载道路图层
var roadUrl=service_ip + (isMicroService ? '/daolu-proxy':'') + '/kqgis/rest/services/daolu/map/';
var serviceUrl=service_ip + (isMicroService ? '/daolu85-proxy':'') + '/kqgis/rest/services/daolu85/networkanalysis';
L.KqGIS.tileMapLayer(roadUrl,
{
style: 'default',
format: 'image/png',
transparent: true
}
).addTo(map);
```

第三步,使用 L.ExtraMarkers.icon 创建图标对象 originIcon,设置其颜色、形状等属性,编号 number 设置为'起',同样地,分别设置 middleIcon、middleIcon2、destinationIcon。使用 L.marker 类创建标注点对象,设置其坐标和图标,使用 addTo 方法将它们加入地图。

```
//标记起点 终点 途经点
letoriginIcon=L.ExtraMarkers.icon({
icon: 'fa-number',
markerColor: 'green',
shape: 'square',
prefix: 'fa',
number: '起'
});
let middleIcon=L.ExtraMarkers.icon({
icon: 'fa-number',
markerColor: 'blue',
shape: 'square',
prefix: 'fa',
number: '途1'
});
letmiddleIcon2=L.ExtraMarkers.icon({
```

```
      icon: 'fa-number',
      markerColor: 'blue',
      shape: 'square',
      prefix: 'fa',
      number: '途 2'
    });
    let destinationIcon=L.ExtraMarkers.icon({
      icon: 'fa-number',
      markerColor: 'blue',
      shape: 'square',
      prefix: 'fa',
      number: '终'
    });
    L.marker([26.670953546577746,106.70700186909606],{ icon: originIcon }).addTo(map);        // 起点
    L.marker([26.647097558343594,106.6987392653254],{ icon: middleIcon }).addTo(map);         //途经点 1
    L.marker([26.637097558343594,106.6947392653254],{ icon: middleIcon2 }).addTo(map);        // 途经点 2
    L.marker([26.609635542110627,106.71528310800609],{ icon: destinationIcon }).addTo(map);   //终点
```

第四步,为 analysis 容器绑定 kendoButton 容器,点击时异步执行函数,首先使用 L.point 初始化起点、终点、途经点等各点信息,然后使用 KqGIS.NetworkAnalysis.BestPathParams 类创建查询参数列表对象。该类参数如表 8-6 所示。

表 8-6　BestPathParams 参数介绍

Name	Type	Default	参数定义
startPt	KqGIS.Geometry.Point | L.Point		起点
destPt	KqGIS.Geometry.Point | L.Point		目标点
midPt	Array.<(KqGIS.Geometry.Point|L.Point)>		可选,途经点
impedance-Mode	KqGIS.ImpedanceMode		可选,最佳路径阻抗模式,支持时间和距离;①ImpedanceLen 长度(最短路径)②ImpedanceTime 时间(最快路径,需要在加载数据的时候指定速度相关参数)

续表 8-6

Name	Type	Default	参数定义
usedData-Mode	string		可选,KQNetworkAnalystDataUsedMode 枚举的字符串表现形式,如果不传,默认为 UsedModeVector;枚举值：UsedModeWithVector：此次分析仅使用矢量；UsedModeWithDemRaster：此次分析仅使用 DEM 栅格；UsedModeMix：混合模式,即使用矢量和栅格
usedRaster-Data	Array.<string>		可选,需要进行栅格分析时,指定具体的栅格数据名称,名称为加载的时候指定的内容;如果没有指定,则使用所有加载的栅格数据进行分析
barrierPt	Array.<(KqGIS.Geometry.Point|L.Point)>		可选,障碍点
attributeFilterOfImpassableRoad	string		可选,属性过滤条件,设置后面分析中不能通行的道路,简单的 sql 数据支持：比如需要限高、限宽的时候
avoid	GeoJSONObject|L.GeoJSON		可选,规避区域
async	boolean	false	可选,是否异步执行分析方法,如果是异步,则直接返回分析任务的 ID,后续可通过该分析任务 ID,取消分析或查询分析结果。默认为同步执行

```
//初始化分析按钮
$('#analysis').kendoButton({
click: async function () {
var startPt=L.point([106.70700186909606,26.670953546577746])
var destPt=newKqGIS.Geometry.Point([106.71528310800609,26.609635542110627]);
var point1=L.point([106.6987392653254,26.647097558343594]);
var point2=L.point([106.6947392653254,26.637097558343594]);
var midPt=[point1,point2];
var params= new KqGIS.NetworkAnalysis.BestPathParams({
startPt: startPt,//起点
destPt: destPt,//终点
```

```
midPt：midPt,//途经点
impedanceMode：KqGIS. ImpedanceMode. ImpedanceLen,
async：false
});
```

第五步,将 serviceUrl、params 作为参数传入,调用 L. KqGIS. networkAnalysisService 的 bestPath 方法进行缓冲分析,设置回调函数。

```
L. KqGIS. networkAnalysisService(serviceUrl). bestPath(params,(serviceResult)
```

如果 serviceResult、serviceResult. result 和 serviceResult. result. resultcod 均为 success,那么首先调用 routeGroup. clearLayers 方法清空已绘制的图层。使用_getLatlngsFromData 方法获取 serviceResult 的经纬度信息,并用 latlngs 保存。将 latlngs 作为参数,使用 L. polyline 创建线要素并使用 addLayer 方法加入 routeGroup。然后_getNodesFromData 获取 serviceResult 的结点信息。使用创建 popupOptions 变量存放标注点样式信息,用 L. icon 创建 nodeIcon 对象来绘制路径上的结点。

```
if (serviceResult && serviceResult. result && serviceResult. result. resultcode && serviceResult. result. resultcode === "success") {
    routeGroup. clearLayers();
    //绘制路径
    let latlngs=_getLatlngsFromData(serviceResult);
    let path=L. polyline(latlngs);
    routeGroup. addLayer(path);
    //绘制结点
    let nodes=_getNodesFromData(serviceResult);
    varpopupOptions={ minWidth: '80',closeButton: false };
    var nodeIcon=L. icon({
        iconUrl: '../../data/images/node. png',
        iconSize: [24,24],// size of the icon
        iconAnchor: [12,12],// point of the icon which will correspond to marker's location
        popupAnchor: [0,-6] // point from which the popup should open relative to the iconAnchor
    });
```

使用 for 循环遍历 nodes 数组,调用 L. marker 类创建标注点对象,设置 nodes 数组中的 coordinate 坐标,icon 为 nodeIcon。以 popupOptions 作为参数,使用_popupTemplateHtml 方法将 nodes 数组的 direction 和 distance 属性作为内文本为该对象绑定 popup 标注。随后用 addLayer 方法将该对象加入 routeGroup 中。若查询不成功,则执行失败回调。

```
for (let i=1; i < nodes.length; ++i) {
routeGroup.addLayer(L.marker(nodes[i].coordinate,{ icon: nodeIcon }).bindPopup
(_popupTemplateHtml(nodes[i].direction,nodes[i].distance),popupOptions));
}} else {
errorNotify;}
},(error) => {
errorNotify;
});}}).data('kendoButton');
//得到结点信息
```

第六步，设置 _getLatlngsFromData 获取经纬度信息函数，创建 latlngs 数组，设置其初值为空。设置 paths 变量，根据 data.result.results.paths 类型决定其存储方式为 JavaScript 对象或是数组。然后遍历 paths，取 paths 中每一个 segments.features 并创建 features 承接。同样地，遍历 features 的 geometry.coordinates 用 coordinates 承接。最后遍历 coordinates，将 coordinates 的一、二维坐标作为经纬度使用 push 方法传入 new 新建的 latlngs 数组中。

```
function _getLatlngsFromData(data) {
let latlngs=[];
let paths=data.result.results.paths ? data.result.results.paths : [];
for (let i=0; i < paths.length; ++i) {
let features=paths[i].segments.features;
for (let j=0; j< features.length; ++j) {
let coordinates=features[j].geometry.coordinates;
for (let k=0; k < coordinates.length; ++k) {
latlngs.push(new L.LatLng(coordinates[k][1],coordinates[k][0]));
}}}
return latlngs;
}
```

第七步，参考上一步构建函数，将 coordinates、direction（features 的 properties.direction 属性）、distance（features 的 properties.length 属性，使用 Number 方法转为数字）传入 nodes 数组。

```
function _getNodesFromData(data) {
let nodes=[];
let paths=data.result.results.paths ? data.result.results.paths : [];
for (let i=0; i < paths.length; ++i) {
let features=paths[i].segments.features;
for (let j=0; j < features.length; ++j) {
```

```
let coordinates=features[j].geometry.coordinates;
nodes.push({
coordinate：[coordinates[0][1],coordinates[0][0]],
direction：features[j].properties.direction,
distance：Number(features[j].properties.length).toFixed(2)
});
}}
return nodes;
}
```

设置弹出信息函数，以 direction、distance 作为参数，设置容器存放"到下一个拐点""direction+方向""distance+米"等信息。

```
//弹出信息模板
function _popupTemplateHtml(direction,distance) {
let elemTips=<divstyle='font-weight：bold; font-size：13px; margin-bottom：4px'>到下一个拐点</div>;
let elemDirection=<div style='margin-bottom：4px'>方向：${direction}</div>;
let elemDistance=<div>距离：${distance}米</div>;
return '<div>' + elemTips + elemDirection + elemDistance + '</div>';
```

设置失败回调函数，为 notification 绑定 kendoNotification 容器，设置样式。使用 setTimeout 方法等待 500 秒后自动调用函数，将 notification 中的信息设置为"分析错误"。

```
setTimeout(function () {
notification.show({
message：'分析错误'
},'info');
},500);
</script>
```

运行效果如图 8-7 所示。

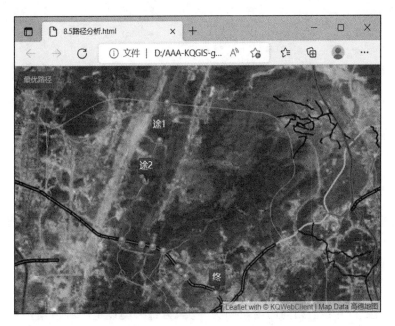

图 8-7 路径分析效果预览

9 KQGIS 专题图

9.1 客户端专题图

客户端专题图用于在 WebGIS 客户端展示某一类或几类数据的统计图,通常不需要调用数据服务,而是采用相关的数据展示包(如 echarts)实现专题图的展示。

9.1.1 客户端单值专题图

单值专题图是将专题值相同的要素归为一类,为每一类设定一种渲染风格,如颜色或符号等,专题值相同的要素采用相同的渲染风格,从而用来区分不同的类别。具体实现方法如下:

第一步,新建 map、infowin、themeLayer、infowinPosition 等变量,并初始化其值为空值。新建 styleGroups 数组,将各种用地类型的名称和对应颜色信息存入数组。

```
<script>
let map=null;
let infowin=null;
let themeLayer=null;
let infowinPosition=null;
let styleGroups=[
{
value:'公路用地',
style:{
fillColor:'#D2D8C9'
}},{
value:'农村居民点用地',
style:{
fillColor:'#EC898A'
}
},{
value:'农村道路',
```

```
style：{
fillColor：'#C2C1C1'
}
},{
value：'园地',
style：{
fillColor：'#E7CCE2'
}
}];
```

第二步,定义 highLightLayer 函数,将鼠标事件作为参数传入。如果鼠标事件的对象存在且有 refDataID 属性,则使用 getFeatureById 函数根据其 refDataID 找到该对象并将其作为结果返回,新建 fea 参数接收返回的对象。如果 fea 存在,则使用 addTo 方法将 infoView 图层加入 map 容器中,并使用 fea 作为参数更新 infoview 图层。如果 fea 不存在且 infoview 存在,则使用 remove 函数将 infoview 移除。

```
function highLightLayer(e) {
if (e.target && e.target.refDataID) {
let fea = themeLayer.getFeatureById(e.target.refDataID);
if (fea) {
infoView.addTo(map);
infoView.update(fea);
}
} else if (infoView) {
infoView.remove();
}
}
```

第三步,定义 handleMapEvent 函数,将地图容器与其他容器作为参数传入,如果二者有一个不存在,则中止运行。否则,使用 addEventListener 为 div 容器添加 mouseover 鼠标悬停事件的监听器,当监听器被触发时,将 map 容器的 dragging、scrollWheelZoom、doubleClickZoom 方法禁用。使用 addEventListener 为 div 容器添加 mouseout 鼠标移出事件的监听器,当监听器被触发时,将 map 容器的 dragging、scrollWheelZoom、doubleClickZoom 方法启用。

```
function handleMapEvent(div,map) {
if (!div||!map) {
return;
}
div.addEventListener('mouseover',function() {
```

```
map.dragging.disable();
map.scrollWheelZoom.disable();
map.doubleClickZoom.disable();
});
div.addEventListener('mouseout',function() {
map.dragging.enable();
map.scrollWheelZoom.enable();
map.doubleClickZoom.enable();
});
}
```

第四步,定义 initLegendView 函数,使用 L.control 自定义 legendView 控件。随后调用 onAdd 函数初始化 legendView 控件,使用 DomUtil 类的 create 方法创建其 div 容器,并设置其基本样式。新建 items 列表,用 for 循环遍历 styleGroups 中的内容,并把 value 和 fillColor 添加进 div 容器中的<tr>标签。

```
//图例控件
function initLegendView() {
let legendView=L.control({ position: 'bottomright' });
legendView.onAdd=function() {
this._div=L.DomUtil.create('div','panel panel-primary legend');
let items='';
for (let i=0; i < styleGroups.length; ++i) {
items +=
`<tr>
<td class='legendItemHeader'>${styleGroups[i].value}</td>
<td class='legendItemValue' style='background: ${styleGroups[i].style.fillColor}'></td>
</tr>`;
}
```

第五步,创建 legendDiv,设置其样式后,使用 appendTo 方法将 legendDiv 添加进 legend-View 中,使用第三步定义 handleMapEvent 函数设置 this._div 和 this._map 容器的鼠标事件,最后使用 addTo 方法将 legendView 加入地图中。

```
let legendDiv=`
<div class='panel-heading'>
<h5 class='panel-title text-center'>图例</h5>
</div><div class='panel-body text-left' ><table><tr>
```

```
<td class='legendItemHeader'>地类名称</td><td class='legendItemValue'>颜色</td>
</tr> ${items}</table></div>`;$(legendDiv).appendTo(this._div);
handleMapEvent(this._div,this._map);return this._div;};
legendView.addTo(map);}
```

第六步,同样地,参照第五步定义 initInfoView 函数,初始化 infoView 控件并添加 content 容器。将容器添加入地图容器,将 infoView 控件添加入地图中。

```
//高亮时显示图层信息框的控件
function initInfoView() {
infoView=L.control({ position: 'topleft' });
infoView.onAdd=function() {
this._div=L.DomUtil.create('div','panel panel-primary infoPane');
$(this._div).css('width','200px');
$(`
<div class='panel-heading'>
<h5 class='panel-title text-center'>属性表</h5>
</div>
`).appendTo(this._div);
let content=$(`<div class='panel-body content'></div>`).appendTo(this._div);
content.css('fontSize','14px');
handleMapEvent(this._div,this._map);
return this._div;
};
```

第七步,使用 update 方法更新 infoview,将 fea 作为参数传入,若查询的 fea 存在,则将 content 中的内容使用 text 方法展示为内文本。若不存在,则退出函数。定义 innerHtml 内容并使用 html 方法,将内文本传入 content。

```
infoView.update=function(fea) {
let content=$('.content');
content.text('');
if (!fea) {
return;}
let innerHtml=`OID:${fea.properties.OID}<br/>标识码:${fea.properties.BSM}<br/>地类名称:${fea.properties.DLMC}<br/>面积:${fea.properties.AreaLength1}<br/>`;
content.html(innerHtml);};}
```

第八步,定义 initThemeLayer 函数,初始化 Unique 单值专题图层,设置其高亮、透明度等效果,通过实例化 ThemeStyle 类来设定图层基础样式及高亮样式。此类将会在后续章节多次使用,其具体参数信息如表 9-1 所示。

表 9-1 ThemeStyle 参数介绍

Name	Type	Default	参数定义
fill	boolean	true	可选,是否填充,不需要填充则设置为 false。如果 fill 与 stroke 同时为 false,将按 fill 与 stroke 的默认值渲染图层
fillColor	string	'#000000'	可选,十六进制填充颜色
fillOpacity	number	1	可选,填充不透明度。取值范围[0,1]
stroke	boolean	false	可选,是否描边,不需要描边则设置为 false。如果 fill 与 stroke 同时为 false,将按 fill 与 stroke 的默认值渲染图层
strokeColor	string	'#000000'	可选,十六进制描边颜色
strokeOpacity	number	1	可选,描边的不透明度。取值范围[0,1]
strokeWidth	number	1	可选,线宽度/描边宽度
strokeLinecap	string	'butt'	可选,线帽样式。strokeLinecap 有 3 种类型:"butt""round""square"
strokeLineJoin	string	'iter'	可选,线段连接样式。strokeLineJoin 有 3 种类型:"miter""round""bevel"
strokeDashstyle	string	'solid'	可选,虚线类型。strokeDashstyle 有 8 种类型:"dot""dash""dashdot""longdash""longdashdot""solid""dashed""dotted"。solid 表示实线
pointRadius	number	6	可选,点半径,单位为像素
shadowBlur	number	0	可选,阴影模糊度(大于 0 有效;)。注:请将 shadowColor 属性与 shadowBlur 属性一起使用来创建阴影
shadowColor	string	'#000000'	可选,阴影颜色。注:请将 shadowColor 属性与 shadowBlur 属性一起使用来创建阴影
shadowOffsetX	number	0	可选,阴影 X 方向偏移值
shadowOffsetY	number	0	可选,阴影 Y 方向偏移值
label	string		专题要素附加文本标签内容
fontColor	string		可选,附加文本字体颜色
fontSize	number	12	可选,附加文本字体大小,单位是像素
fontStyle	string	'normal'	可选,附加文本字体样式。可设值:"normal""italic""oblique"
fontVariant	string	'normal'	可选,附加文本字体变体。可设值:"normal""small-caps"

续表 9-1

Name	Type	Default	参数定义
fontWeight	string	'normal'	可选，附加文本字体粗细。可设值："normal""bold""bolder""lighter"
fontFamily	string	'arial, sans-serif'	可选，附加文本字体系列。fontFamily 值是字体族名称或/及类族名称的一个优先表，每个值逗号分隔，浏览器会使用它可识别的第一个可以使用具体的字体名称（"times""courier""arial"）或字体系列名称（"serif""sans-serif""cursive""fantasy""monospace"）
labelPosition	string	'top'	可选，附加文本位置，可以是 'inside' 'left' 'right' 'top' 'bottom'
labelAlign	string	'center'	可选，附加文本水平对齐。可以是 'left' 'right' 'center'
labelBaseline	string	'middle'	可选，附加文本垂直对齐。可以是 'top' 'bottom' 'middle'
labelXOffset	number	0	可选，附加文本在 x 轴方向的偏移量
labelYOffset	number	0	可选，附加文本在 y 轴方向的偏移量

```
function initThemeLayer() {
//定义 Unique 单值专题图层
themeLayer = L. KqGIS. uniqueThemeLayer('ThemeLayer',{
//开启 hover 高亮效果
isHoverAble: true,
opacity: 0.8,
alwaysMapCRS: true
}).addTo(map);
//图层基础样式
themeLayer. style = new KqGIS. ThemeStyle({
shadowBlur: 3,
shadowColor: '#000000',
shadowOffsetX: 1,
shadowOffsetY: 1,
fillColor: '#FFFFFF'
});
// hover 高亮样式
themeLayer. highlightStyle = new KqGIS. ThemeStyle({
stroke: true,
strokeWidth: 2,
strokeColor: 'blue',
```

```
fillColor:'#00F5FF',
fillOpacity:0.2
});
```

第九步,设置专题图属性字段名称及风格数组,添加专题图层的 mousemove 事件,当鼠标悬停在高光图层上方时,触发事件。定义 addThemeLayer 函数,当调用该函数时,会将 thereData 加入 themeLayer 专题图层中,并调用之前定义的 initLegendView、initInfoView 函数。

```
// 用于单值专题图的属性字段名称
themeLayer.themeField='DLMC';
//风格数组,设定值对应的样式
themeLayer.styleGroups=styleGroups;
themeLayer.on('mousemove',highLightLayer);
addThemeLayer();}
function addThemeLayer() {
themeLayer.addFeatures(thereData);initLegendView();initInfoView();
}
```

第十步,参照 3.1 创建地图容器并初始化地图,设置地图投影中心、缩放层级、地图边界等属性。使用 on 函数注册 mousemove 鼠标事件,当鼠标移入时,将 infowinPosition 设置为鼠标所在图层的点,最后添加,tiandituTilelayer 的矢量图层及矢量注记图层作为底图,调用 initThemeLayer 展示专题图。

运行效果如图 9-1 所示。

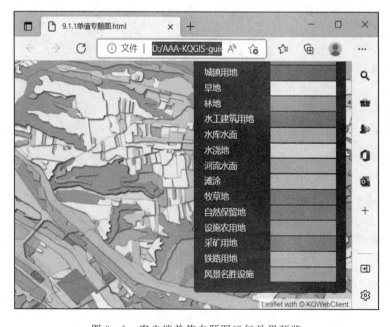

图 9-1　客户端单值专题图运行效果预览

9.1.2　客户端分段专题图

分段专题图是利用图层的某一字段属性,将属性值划分为不同的连续段落(分段范围),每一段落使用不同的符号(线型、填充或者颜色)表示该属性字段的整体分布情况,从而体现属性值和对象区域的关系。具体实现方法如下。

第一步,新建 map 并初始化其值为空值。新建 styleGroups 数组,将属性范围的起始值和终止值、对应颜色信息等存入数组。

```
let map=null;
let styleGroups=[{
start:0,
end:10,
style:{
color:'#FFEDA0'
}},{start:10,end:20,
style:{color:'#FED976'}},{start:20,end:50,
style:{color:'#FEB24C'}},{start:50,end:100,
style:{color:'#FD8D3C'}}];
```

第二步,参照 3.1 创建地图容器并初始化地图,设置地图投影中心、缩放层级、地图边界等属性。使用 on 函数注册 mousemove 鼠标事件,当鼠标移入时,将 infowinPosition 设置为鼠标所在图层的点,最后添加,tiandituTilelayer 的矢量图层及矢量注记图层作为底图,调用 initThemeLayer 函数展示专题图。

第三步,定义 initThemeLayer 函数,初始化 rangeThemeLayer 单值专题图层,设置其高亮、透明度等效果,通过实例化 ThemeStyle 类来设定图层基础样式及高亮样式。

第四步,设置专题图属性字段名称及风格数组,添加专题图层的 mousemove 事件,当鼠标悬停在高光图层上方时,触发事件。定义 addThemeLayer 函数,当调用该函数时会将 density 加入 themeLayer 专题图层中,并调用 initLegendView(用于显示图例控件)、initInfoView 函数(用于高亮时显示图层信息框),具体定义方法参考 9.1。

```
let innerHtml=`
FID:${fea.properties.FID}<br/>
行政区名:${fea.properties.NAME}<br/>
面积:${(fea.properties.AreaLength1 / 1000000).toFixed(2)} <br/>
人口密度:${fea.properties.DENSITY}<br/>`;content.html(innerHtml);});}
//图例控件
function initLegendView() {
let legendView=L.control({ position:'bottomright' });
```

```javascript
legendView.onAdd=function(){
this._div=L.DomUtil.create('div','panel panel-primary legend');
let items='';
for(let i=0;i<styleGroups.length;++i){
if(i==styleGroups.length-1){
items+=
`<tr>
<td class='legendItemHeader'>${styleGroups[i].start}+</td>
<td class='legendItemValue' style='background:${styleGroups[i].style.color}'></td>
</tr>`;
}else{
items+=
`<tr>
<td class='legendItemHeader'>${styleGroups[i].start}-${styleGroups[i].end}</td>
<td class='legendItemValue' style='background:${styleGroups[i].style.color}'></td>
</tr>`;
}
}
let legendDiv=`
<div class='panel-heading'>
<h5 class='panel-title text-center'>图例</h5>
</div>
<div class='panel-body text-left'>
<table>
<tr>
<td class='legendItemHeader'>2018年人口密度</td>
<td class='legendItemValue'>颜色</td>
</tr>
${items}
</table>
</div>
`;
$(legendDiv).appendTo(this._div);
```

```
handleMapEvent(this._div,this._map);
return this._div;
};
legendView.addTo(map);
}
```

第五步,定义 highLightLayer 函数,将鼠标事件作为参数传入。如果鼠标事件的对象存在且有 refDataID 属性,则使用 getFeatureById 函数根据其 refDataID 找到该对象并将其作为结果返回,新建 fea 参数接收返回的对象。如果 fea 存在,则使用 addTo 方法将 infoView 图层加入 map 容器中,并使用 fea 作为参数更新 infoview 图层。如果 fea 不存在且 infoview 存在,则使用 remove 函数将 infoview 移除。

```
function highLightLayer(e) {
if (e.target && e.target.refDataID) {
var fea=themeLayer.getFeatureById(e.target.refDataID);
if (fea) {infoView.addTo(map);infoView.update(fea);}} else if (infoView) {infoView.remove();}}
```

第六步,定义 handleMapEvent 函数,将地图容器与其他容器作为参数传入,如果二者有一个不存在,则中止运行。否则,使用 addEventListener 为 div 容器添加 mouseover 鼠标悬停事件的监听器,当监听器被触发时,将 map 容器的 dragging、scrollWheelZoom、doubleClickZoom 方法禁用。使用 addEventListener 为 div 容器添加 mouseout 鼠标移出事件的监听器,当监听器被触发时,将 map 容器的 dragging、scrollWheelZoom、doubleClickZoom 方法启用。

```
function handleMapEvent(div,map) {
if (!div||!map) {
return;
}
div.addEventListener('mouseover',function() {
map.dragging.disable();
map.scrollWheelZoom.disable();
map.doubleClickZoom.disable();});
div.addEventListener('mouseout',function() {
map.dragging.enable();
map.scrollWheelZoom.enable();
map.doubleClickZoom.enable();
});
}
</script>
```

运行效果如图 9-2 所示。

图 9-2　客户端分段专题图运行效果预览

9.1.3　客户端等级符号专题图

等级符号专题图与分段专题图类似,同样将矢量图层的某一属性字段信息映射为不同等级,每一级分别使用大小不同的点符号表示,符号的大小与该属性字段值成比例,属性值越大,专题图上的点符号就越大,反之亦同。等级符号专题图多用于具有数量特征的地图上,例如不同地区的粮食产量、GDP、人口等的分级。

第一步,初始化变量,创建 clearThemeLayer 函数清除专题图层内容。当调用该函数时,首先使用 themeLayer.clear()清除图层内容,然后使用 closeInfoWin()关闭弹窗。

```
<script>
let map=null;
let infowin=null;
let themeLayer=null;
let infowinPosition=null;
//清除专题图层中的内容
function clearThemeLayer() {
    themeLayer.clear();
    closeInfoWin();
}
```

创建 showInfoWin 显示地图弹窗。当调用该函数时，先用 if 语句判断鼠标事件指向的对象是否存在，若对象存在，且 target.refDataID 和 target.dataInfo 属性也都存在。先关闭弹窗，使用 themeLayer.getFeatureById 返回 target.refDataID 对象并创建变量 fea 存储。如果 fea 不存在，则直接返回。否则，创建 info 变量储存 target.dataInfo。

```
//显示地图弹窗
function showInfoWin(e) {
// e.target 是图形对象，即数据的可视化对象。
//图形对象的 refDataID 属性是数据（feature）的 id 属性，它指明图形对象由哪个数据制作而来；
//图形对象的 dataInfo 属性是图形对象表示的具体数据，它有 3 个属性：field、R 和 value；
if (e.target && e.target.refDataID && e.target.dataInfo) {
closeInfoWin();
//获取图形对应的数据（feature）
var fea=themeLayer.getFeatureById(e.target.refDataID);
if (! fea) {
return;
}
var info=e.target.dataInfo;
```

设置弹窗内文本样式及默认显示内容 contentHTML，然后根据 info.field 的值选择内文本的显示内容。使用 map.layerPointToLatLng 方法提取 infowinPosition 的坐标点并使用 latLng 储存。若此时 infowin 弹窗不存在，则使用 L.popup 创建弹窗对象 infowin。将 latlng、contentHTML、map 作为参数初始化 infowin。

```
//弹窗内容
var contentHTML='<div style=\'color: #000; background-color: #fff\'>';
contentHTML += '省级行政区名称:' + '<br><strong>' + fea.attributes.NAME + '</strong>';
contentHTML += '<hr style=\'margin: 3px\'>';
switch (info.field) {
case 'CON2009':
contentHTML += '09 年居民消费水平' + ' <br/><strong>' + info.value + '</strong>(' + '元)';
break;
default:
contentHTML += 'No Data';
}
```

```
contentHTML += '</div>';
var latLng=map.layerPointToLatLng(infowinPosition);
if (! infowin) {
infowin=L.popup();
}
infowin.setLatLng(latLng);
infowin.setContent(contentHTML);
infowin.openOn(map);
}
}
```

第二步,创建 closeInfoWin 函数用于移除和销毁地图弹窗,当 infowin 存在时,使用 try 语句,用 remove 方法对 infowin 进行移除。若抛出异常,则使用 alert 将异常信息打印出来。创建 initThemeLayer 初始化专题图层,首先使用 L.KqGIS.rankSymbolThemeLayer 类创建 themelayer 专题图对象,该类的参数除了 name 专题图层名、symbolType 符号类型外,还包括 options 参数,其值如表 9-2 所示。

表 9-2 rankSymbolThemeLayer 参数介绍

Name	Type	Default	参数定义
isOverLay	boolean	true	可选,是否进行压盖处理,如果设为 true,图表绘制过程中将隐藏对已在图层中绘制的图表产生压盖的图表
themeFields	string		指定创建专题图字段
alwaysMapCRS	boolean	false	可选,要素坐标是否和地图坐标系一致,要素默认是经纬度坐标
id	string		可选,专题图层 ID。默认使用 CommonUtil.createUniqueID ("themeLayer_") 创建专题图层 ID
opacity	number	1	可选,图层透明度
TFEvents	Array		可选,专题要素事件临时存储
attribution	string		可选,版权描述信息

```
//移除和销毁地图弹窗
function closeInfoWin() {
if (infowin) {
try {
infowin.remove();
} catch (e) {
alert(e.message);
```

```
}}}
functioninitThemeLayer() {
themeLayer=L.KqGIS.rankSymbolThemeLayer('themeLayer','Circle');
//指定用于专题图制作的属性字段  详看下面 addThemeLayer()中的 feature.attrs.CON2009
themeLayer.themeField='CON2009';
//配置图表参数
themeLayer.symbolSetting={
//允许图形展示的值域范围,此范围外的数据将不制作图形,必设参数
codomain:[0,40000],
//圆最大半径默认100
maxR:100,
//圆最小半径默认0
minR:0,
//圆形样式
circleStyle:{ fillOpacity:0.7 },
//符号专题图填充颜色
fillColor:'#0E10C5',
//专题图 hover 样式
circleHoverStyle:{ fillOpacity:0.8 }
};
```

配置完专题图参数后,使用 addTo 方法把 themeLayer 加入地图。然后为专题图添加 on 鼠标事件,当鼠标移入,调用 showInfoWin 显示专题图,反之关闭专题图。然后调用 addThemeFeatures 添加 features 数据,最后使用 initInfoView 初始化信息显示。

```
themeLayer.addTo(map);
//注册专题图 mousemove,mouseout 事件(注意:专题图图层对象自带 on 函数,没有 events 对象)
themeLayer.on('mousemove',showInfoWin);
themeLayer.on('mouseout',closeInfoWin);
addThemeFeatures();
initInfoView();
}
```

第三步,创建 addThemeFeatures 函数,该函数被调用时,创建 features 数组,使用 for 循环遍历 chinaConsumptionLevel.js 文件中的数据,创建 provinceInfo 存储该文件中某一个省的数据。读取 provinceInfo 的第二三位数据,并使用 L.point 创建点要素,命名为 geo。创建 attrs 变量,取 provinceInfo 的第一位数据作为 attrs 变量的 NAME 属性,取第四位数据作为

CON2009 属性。以 geo 和 attrs 作为参数，使用 L. KqGIS. themeFeature 创建 feature 对象。使用 push 方法把 feature 压入数组 features 中，使用 addFeatures 方法把 features 数组加入 themeLayer 数组中。

```
//构建 feature 数据
function addThemeFeatures() {
let features=[];
for (let i=0,len=chinaConsumptionLevel. length; i < len;i++) {
//省居民消费水平(单位:元)信息
let provinceInfo=chinaConsumptionLevel[i];
let geo=L. point(provinceInfo[1],provinceInfo[2]);
let attrs={ NAME：provinceInfo[0],CON2009：provinceInfo[3] };
let feature=L. KqGIS. themeFeature(geo,attrs);
features. push(feature);
}
themeLayer. addFeatures(features);
}
```

创建 initInfoView 函数，当该函数被调用时。使用 L. control 创建自定义控件 legendView 并设置其位置为 topleft 左上角。使用 onAdd 为控件添加函数，当该函数被调用时，使用 L. DomUtil. create 创建名为 info 的容器，将其与 this._div 对象绑定，设置对象的 text 属性为"09 年城市居民消费图"。返回后，使用 addTo 方法加入地图。

```
function initInfoView() {
let legendView=L. control({ position：'topleft' });
legendView. onAdd=function() {
this._div=L. DomUtil. create('div','info');
$(this._div). text("09 年城市居民消费图")
return this._div;
};
legendView. addTo(map);
}
```

第四步，完成地图初始化及容器初始化，并将它们用 addTo 方法加入地图中。

调用 initThemeLayer 函数初始化专题图，为 map 添加鼠标事件，当鼠标移入时，读取当前的 layerPoint 并使用 infowinPosition 存储。

```
initThemeLayer();
map. on('mousemove',function(e) {
infowinPosition=e. layerPoint;
```

```
});
}
</script>
```

运行效果如图 9-3 所示。

图 9-3 客户端等级符号专题图运行效果

9.1.4 客户端 Echarts 统计图

EChars 图属于动态统计图,具有动画效果,并提供工具条,可对统计图进行各种操作,如添加辅助线、切换统计图、查看数据视图、保存为图片等。具体实现方式如下。

第一步,完成地图初始化及容器初始化,并用 addTo 方法将它们加入地图。

第二步,创建 data 数组,存放各个城市的经纬度坐标信息。创建变量 convertData 并绑定函数,当该函数被调用时,先创建空数组 res,然后使用 for 循环遍历 data 中每一个元素。使用 geoCoordMap 存放元素的 name,用 geoCoord 承接。若 geoCoord 存在,则把元素的 name 属性使用 concat 函数进行处理,然后使用 push 方法将处理结果的 value 属性存入数组 res 中,最后返回 res 数组。

```
var data=[
'廊坊': [116.7,39.53],
'菏泽': [115.480656,35.23375],
'合肥': [117.27,31.86],
```

```
'武汉': [114.31,30.52],
'大庆': [125.03,46.58]
};
var convertData=function (data) {
var res=[];
for (var i=0; i<data.length; i++) {
var geoCoord=geoCoordMap[data[i].name];
if (geoCoord) {
res.push({
name: data[i].name,
value: geoCoord.concat(data[i].value)
});
}
}
return res;
};
```

第三步,创建专题图 options 列表,设置 title 标题、toollip 工具条(触发条件 trigger 为 item)、legend 文本显示域的样式和数据来源。

```
option= {
title: {
text: "全国主要城市空气质量图",
subtext: 'data from PM25.in',
sublink: 'http://www.pm25.in',
left: 'center',
top: 20,
textStyle: {
color: '#fff'
}},
tooltip: {
trigger: 'item'
},
legend: {
orient: 'vertical',
y: 'bottom',
x: 'right',
data: ['pm2.5'],
```

```
textStyle: {
color: '#fff'
}},
```

设置 series,创建 name 为 pm2.5 的数据集,设置其 type 为 scatter 散点图,设置坐标系统 coordinateSystem 为 leaflet。Data 为经过 convertData 转换后的 data 数组。symbolSize 方法设置为传入 val 数组后,取第三位的值,将其除以 10 返回。设置 label、emphasis 等效果的样式信息。

```
series: [
{
name: 'pm2.5',
type: 'scatter',
coordinateSystem: 'leaflet',
data: convertData(data),
symbolSize: function (val) {
return val[2] / 10;
},
label: {
normal: {
formatter: '{b}',
position: 'right',
show: false
},
emphasis: {
show: true
}
},
itemStyle: {
normal: {
color: '#ddb926'
}}},
```

同样地,设置名为 Top 5 的数据集及其他参数属性。

```
{
name: 'Top 5',
type: 'effectScatter',
coordinateSystem: 'leaflet',
```

```
data: convertData(data.sort(function (a,b) {
return b.value - a.value;
}).slice(0,6)),
symbolSize: function (val) {
return val[2] / 10;
},
showEffectOn:'render',
rippleEffect: {
brushType: 'stroke'
},
hoverAnimation: true,
label: {
normal: {
formatter: '{b}',
position: 'right',
show: true
}},
itemStyle: {
normal: {
color: '#f4e925',
shadowBlur: 10,
shadowColor: '#333'
}},
zlevel: 1
}]);
```

第四步,使用 L.KqGIS.echartsLayer 类创建 echarts 专题图层,并用 addTo 方法加入地图。该类参数如表 9-3 所示。

表 9-3 echartsLayer 参数介绍

Name	Type	Default	参数定义
loadWhileAnimating	boolean	false	可选,是否在移动时实时绘制
attribution	string	'© 2018 百度 ECharts'	可选,版权信息

```
L.KqGIS.echartsLayer(option).addTo(map);
}
</script>
```

运行效果如图 9－4 所示。

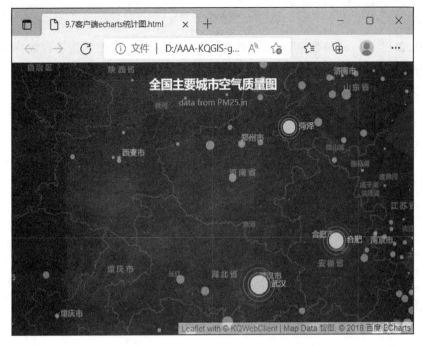

图 9－4　客户端 echarts 统计图效果预览

9.1.5　客户端热力图

热力图是一种通过对色块着色来显示数据的统计图表。绘图时,需指定颜色映射的规则。例如,较大的值由较深的颜色表示,较小的值由较浅的颜色表示;较大的值由偏暖的颜色表示,较小的值由较冷的颜色表示。具体实现方式如下。

第一步,创建变量及图层切换函数,当传入 value 时,先调用 clearLayers 清除所有图层,然后根据 value 的值将对应图层加入 layergroup 图层组中。

```
<script>
varcommomColorLayer,customColorLayer,map,layergroup;
function changeLayers(value) {
layergroup.clearLayers();
if (value === 'common') {
commomColorLayer.addTo(layergroup);
} else if (value === 'custom_color') {
customColorLayer.addTo(layergroup);}}
```

创建 onChange 函数,当该函数被调用时,使用 val 方法取 selector 容器的值,并赋给 value 变量。然后以 value 为参数调用 changeLayers 函数。

```
function onChange() {
var value=$("#selector").val();changeLayers(value);};
```

第二步,创建 onload 函数,当该函数被调用时,先创建 data 数组,用于存放待选的选项文本和数值信息。为 selector 容器绑定 kendoDropDownList,设置其文本域为 text、值域为 value、数据源为 data,Index 为 0、change 属性为 onChange 函数。

```
function onload() {
var data=[
{ text:"热力图",value:"common" },
{ text:"自定义颜色",value:"custom_color" },];
$("#selector").kendoDropDownList({dataTextField:"text",dataValueField:"value",dataSource:data,index:0,
change:onChange
});
```

第三步,完成地图初始化及容器初始化,并用 addTo 方法将它们加入地图。

定义热力点数量 heatNumbers、热力半径 heatRadius、热力坐标信息数组 heatPoints,以【39.78,116.12】为中心使用 Math.random()遍历添加随机值到 heatPoints 数组中。

```
var heatNumbers=150,heatRadius=30;
var heatPoints=[];
for (var i=0; i<heatNumbers; i++) {
heatPoints[i]=[Math.random() * 0.28 + 39.78,Math.random() * 0.5 + 116.12,
Math.random() * 80];
}
```

以 heatPoints 为参数,使用 L.heatLayer 创建热力图层 commomColorLayer,设置其半径为 heatRadius、最小透明度为 0.5。

```
//普通的热力图图层
commomColorLayer=L.heatLayer(heatPoints,{
radius:heatRadius,
minOpacity:0.5,
});
```

同样地,创建 customColorLayer,使用 changeLayers 方法设置 common 为默认图层。

```
//自定义颜色的热力图图层
customColorLayer=L.heatLayer(heatPoints,{
```

```
        radius: heatRadius,
        minOpacity: 0.5,
        gradient: { 0.4: 'blue', 0.65: 'lime', 1: 'red' }
    });
    changeLayers('common');
}
</script>
```

运行效果如图 9-5 所示。

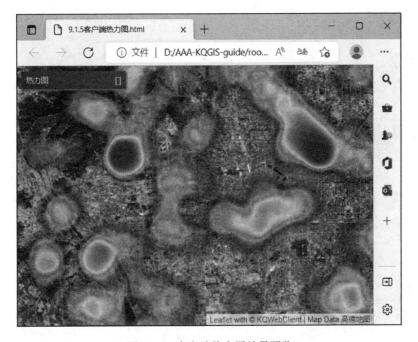

图 9-5　客户端热力图效果预览

9.2　专题图服务

专题图服务是已经被上传且封装好的专题图生成接口，通过调用该接口可以为数据生成各类专题图，以下列几个专题图为例。

9.2.1　四色专题图服务

行政区划图着色要求相邻行政区划单元颜色不相同，在计算机领域通常被称为"四色图"问题，即对于任一地图，只用不多于 4 种颜色，就能使相邻的行政区划单元颜色不相同。具体实现方式如下。

第一步，创建变量 map、maplayer、函数 onload，当该函数被调用时，创建地图服务地址 serviceUrl。然后完成地图初始化及容器初始化，并用 addTo 方法将它们加入地图。

第二步，为 notification 容器绑定 kendoNotification 消息展示框设置其样式及属性。

```
var notification = $("#notification").kendoNotification({
position: {
pinned: true,
bottom: 12,
right: 12
},
autoHideAfter: 2000,
stacking: "up",
templates: [{
type: "info",
template: <div>#= message #</div>
}]
}).data("kendoNotification");
```

创建_params 变量，设置其 layer 属性，包括 type、name、refdataset、theme、colors 等。其中 type 的类型为 KqGIS.ThematicType.FOURCOLOR。在该命名空间下，还有以下参数，如表 9-4 所示。

表 9-4 ThematicType 命名空间参数介绍

Name	Type	Default	参数定义
SINGLE	string	single	单一渲染专题图
DOTDENSITY	string	dotdensity	点密度专题图
CLUSTER	string	cluster	格网聚合专题图
UNIQUE	string	unique	唯一值专题图
RANGE	string	range	分段专题图
PIE	string	pie	统计专题——饼状图
HISTOGRAM	string	histogram	统计专题——直方图
STACK	string	stack	统计专题——堆叠图
LEVEL	string	level-symbol	分级符号
FOURCOLOR	string	fourColor	四色图
LABEL	string	label	标签图

```
var _params={
"layer": {
"type": "featurelayer",
"name": "四色图专题图",
"refdataset": "LANDUSE_R",
"theme": {
"type":KqGIS.ThematicType.FOURCOLOR,
"colors": [{"red":255},{"green":255},{"blue":255},{"red":100}]}}}
```

用 formatJSON 将 _params 转为 JSON 字符,并使用.val 将值赋给 params 容器。为 query 容器绑定按钮,当按钮被按下时,设置 result 容器的 val 为空值,设置 loading 容器的 css 样式为 display/block。最后调用 createTheme 函数。

```
$("#params").val(formatJSON(_params));
$('#query').click(function () {
$("#result").val('');
$("#loading").css("display","block");
createTheme()})
```

第三步,创建 createTheme 函数,当该函数被调用时,传入 serviceUrl 使用 L.KqGIS.mapThemeService 创建地图专题图服务对象 mapThemeService。使用 JSON.parse 方法把 params 的 val 值转为 JavaScript 对象,用 params 变量承接。使用 Object.assign 将所有可枚举的自有属性从一个或多个源对象复制到目标对象,返回修改后的对象。这里的源对象是 params.layer,目标对象使用 new KqGIS.Map.FourColorTheme 创建的对象 theme。其 type 为源对象的 theme.type 属性。遍历源对象 theme.colors 属性生成的数组 item,使用 push 方法将 item 中的值压入 result 中。

```
functioncreateTheme() {
var mapThemeService=L.KqGIS.mapThemeService(serviceUrl)
var params=JSON.parse($("#params").val())
Object.assign(params.layer,{
theme: new KqGIS.Map.FourColorTheme({
type: params.layer.theme.type,
colors: (function () {
var result=[]
for (var i=0,item; item=params.layer.theme.colors[i++];) {
item=new KqGIS.Color(item)
result.push(item)
}
```

第四步，使用 KqGIS.Map.CreateThemeParams 创建查询列表对象 params。

params=new KqGIS.Map.CreateThemeParams(params)

然后使用 mapThemeService.create 调用专题图创建，若成功调用服务，则设置 loading 容器的 css 样式为 display/block。设置 notification 中的信息为"查询成功"。创建 response 对象承接查询结果的 result 属性。如果 response.resultcode 的值不为 success，则直接返回。否则，继续执行函数。若当前存在 maplayer 图层，则将其移除，否则以 serviceUrl 为参数使用 L.KqGIS.tileMapLayer 创建 mapLayer，其 layerid 属性为 response.result。使用 addTo 方法加入地图。将 response 对象转为 JSON 后，使用 val 方法赋值给 result 容器进行显示。若抛出异常，则设置 loading 容器的 css 样式为 display/block。设置 notification 中的信息为"查询失败"。

```
mapThemeService.create(params,(response) => {
$("#loading").css("display","none");
notification.show({
message:"查询成功!"
},"info");
response=response.result
if (response.resultcode ! == 'success') {
return;}
if (mapLayer) {
mapLayer.remove()}mapLayer=L.KqGIS.tileMapLayer(serviceUrl,
{layerIds:response.result}).addTo(map)
$("#result").val(formatJSON(response));},(error) => {
$("#loading").css("display","none");
notification.show({message:"查询失败!"},"info");
})
```

运行效果如图 9-6 所示。

9.2.2 统计专题图

统计专题图是根据地图属性表中所包含的统计数据进行制图，可在地图中形象地反映同一类属性字段之间的关系。借助统计专题图可以更好地分析自然现象和社会经济现象的分布特征和发展趋势，例如研究区植被类型分布变化或城市人口增长比率。

这里以直方统计图为例，具体实现方式与 9.2.1 中基本一致。但在_params 变量的设置中，要将 type 的类型设置为 KqGIS.ThematicType.HISTOGRAM。

9　KQGIS 专题图

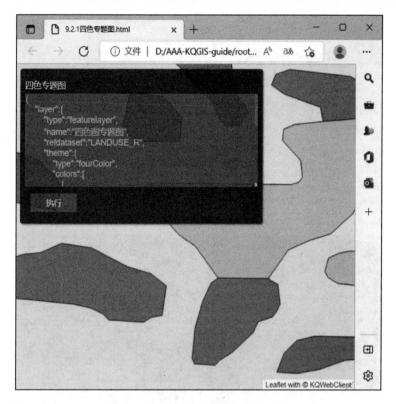

图 9-6　四色专题图服务效果预览

```
var _params={
"layer": {
"type": "featurelayer","name": "直方图专题图","refdataset": "LANDUSE_R",
"theme": {
"type": KqGIS.ThematicType.HISTOGRAM,"radius": "10px","items": [{
"field": "Area","color": {"red": 0,"green": 255,"blue": 0,"alpha": 255}},{
"field": "Area_1","color": {"red": 255,"green": 0,"blue": 0,"alpha": 255}},
]}}}
```

在 Object.assign 中,使用 KqGIS.Map.HistogramTheme 来创建专题图对象。

```
Object.assign(params.layer,{
theme: new KqGIS.Map.PieTheme({
type: params.layer.theme.type,
radius: params.layer.theme.radius,
items: (function () {var result=[]
for (var i=0,item; item=params.layer.theme.items[i++];) {
```

```
item.color=new KqGIS.Color(item.color)
result.push(newKqGIS.Map.StatisticItem(item)}
return result
})()
})
```

运行效果如图 9-7 所示。

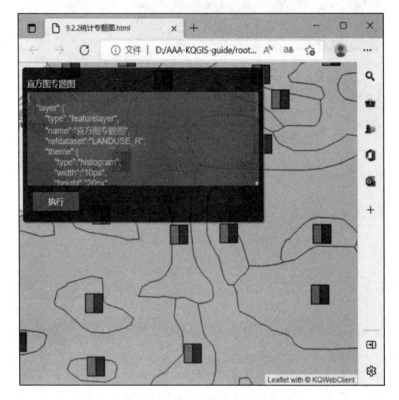

图 9-7 统计专题图服务效果预览

9.2.3 点密度专题图

点密度专题图与分段专题图和等级符号专题图类似,同样将矢量图层的某一属性字段信息映射为不同等级,每一级别使用表现为密度形式的点符号表示,点符号分布在区域内的密度高低与该属性字段值成比例,属性值越大,专题图上的点符号的分布就更为密集,反之亦同。

具体实现方式与 9.2.1 基本一致,但在_params 变量的设置中,要将 type 的类型设置为 KqGIS.ThematicType.DOTDENSITY。

```
var _params={
"layer": {
"type": "featurelayer",
"name": "点密度_1",
"refdataset": "Neighbor_R",
"theme": {
"type": KqGIS.ThematicType.DOTDENSITY,
"dotsize": "2mm",
"dotvalue": 5000,
"items": [{"expression": "ADMI","color":{"red": 255,"green": 0,"blue": 0,"alpha": 255
}},
]}}}
```

在 Object.assign 中,使用 KqGIS.Map.DotdensityTheme 来创建专题图对象。

```
Object.assign(params.layer,{
theme: new KqGIS.Map.DotdensityTheme({type:params.layer.theme.type,
dotsize:params.layer.theme.dotsize,dotvalue:params.layer.theme.dotvalue,
items:(function(){var result=[]for(var i=0,item;item=params.layer.theme.items[i++];){
item.color=new KqGIS.Color(item.color)result.push(new KqGIS.Map.DotdensityItem(item))}
return result
})()
```

运行效果如图 9-8 所示。

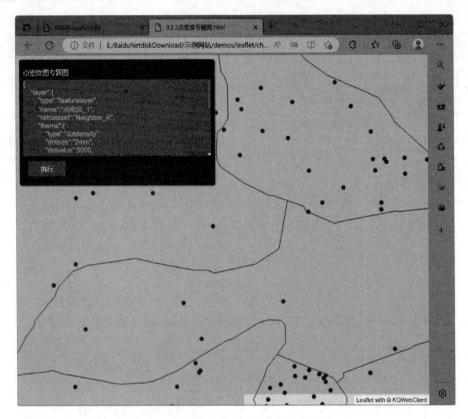

图 9-8 点密度专题图服务效果预览

主要参考文献

郭明强,2023.WebGIS 之 Cesium 三维软件开发[M].北京:电子工业出版社.

郭明强,黄颖,2019.WebGIS 之 OpenLayers 全面解析[M].2 版.北京:电子工业出版社.

郭明强,黄颖,2021.WebGIS 之 Leaflet 全面解析[M].北京:电子工业出版社.

郭明强,黄颖,李婷婷,等,2021.WebGIS 之 Element 前端组件开发[M].北京:电子工业出版社.

郭明强,黄颖,潘雄,等,2022.地理空间信息系统设计与开发[M].武汉:中国地质大学出版社.

郭明强,黄颖,容东林,2022.网络地理信息服务开发实践[M].武汉:中国地质大学出版社.

郭明强,黄颖,万晓明,等,2022.互联网 GIS 系统开发实践[M].武汉:中国地质大学出版社.

郭明强,黄颖,吴亮,等,2022.移动 GIS 应用开发实践[M].北京:电子工业出版社.

郭明强,黄颖,杨亚仑,等,2021.WebGIS 之 ECharts 大数据图形可视化[M].北京:电子工业出版社.

吴信才,吴亮,万波,等,2020.地理信息系统应用与实践[M].北京:电子工业出版社.

吴信才,吴亮,万波,等,2022.地理信息系统应用与实践[M].2 版.北京:电子工业出版社.